PLANT DISEASES
of
Viral, Viroid, Mycoplasma and Uncertain Etiology

T0133929

Editor
KARL MARAMOROSCH

Routledge
Taylor & Francis Group

NEW YORK AND LONDON

First published 1992 by Westview Press

Published 2019 by Routledge
605 Third Avenue, New York, NY 10017
2 Park Square, Milton Park, Abingdon, Oxon OX14 4RN

First issued in paperback 2022

Routledge is an imprint of the Taylor & Francis Group, an informa business

Publisher's Note
The publisher has gone to great lengths to ensure the quality of this reprint but points out that some imperfections in the original copies may be apparent.

CIP data available upon request

ISBN 13: 978-0-367-29848-7 (pbk)
ISBN 13: 978-0-367-28302-5 (hbk)

CONTENTS

PREFACE

The causes of several important plant diseases are still unknown or uncertain. Economically important food and fiber crops as well as trees in different parts of the world are affected by pathogens the nature of which has not yet been well established. Considerable benefits can be expected from the study of the nature of such diseases because this can result in the devising of proper control methods.

In order to focus attention at the as yet uncertain etiology diseases of plants a joint Indo-U.S. workshop has been organized and held at the Indian Agricultural Research Institute in New Delhi in November, 1989. Karl Maramorosch, the Editor of the present volume was the Project Director of this meeting and the Division of International Programs of the National Science Foundation in Washington, D.C. supported the workshop with a grant to Rutgers—The State University of New Jersey. It was felt that the potential for joint projects to develop between scientists of India and other countries would be significant in areas of basic understanding and practical methods of plant disease control. Following the transfer of information meeting, the U.S. National Science Foundation has agreed to use funds from their international grant to subsidize the publication of selected contributions of workshop participants and invited authors from the United States and other countries. Additional contributions dealing exclusively with viroids, by authors from the United States, Australia, Canada, China and Italy have been edited by Karl Maramorosch and published by CRC Press in the United States in a volume entitled "Viroids and Satellites: Molecular Pathogens at the Frontier of Life". Abstracts of the November 15-18, 1989 Indo-U.S. Workshop on Viroids and Diseases of Uncertain Etiology have been published by the Advanced Centre for Plant Virology, Indian Agricultural Research Institute, New Delhi 110 012, with funds provided by the Far Eastern Regional Research Office of the U.S. Department of Agriculture, as well as Indian Agricultural Research Institute and ICAR.

The contributors of this book are active workers from the United States, Canada, United Kingdom, Germany, France, Poland and Brazil. The subject of this treatise will appeal to plant pathologists, entomologists, horticulturists and forest pathologists, as well as others interested in different branches of agriculture.

KARL MARAMOROSCH
Editor

THE CONTRIBUTORS

E. Beuther
Institut für Physikalische Biologie
Heinrich-Heine Universität Düsseldorf
D-4000 Düsseldorf, Germany

Luigi Chiarappa
4221 Montgomery Avenue
Davis, California 95616, USA

Michel Dolet
IRHO Virology Division
Laboratoire de Phytovirologie des Regions Chaudes
CIRAD-ORSTOM
BP 5035
34032 Montpellier Cedex 1, France

Hugh C. Harries
17 Alexandra Road
Lodmoor Hill, Weymouth
Dorset DT4 7QQ, England, UK

Chuji Hiruki
Department of Plant Science
University of Alberta
Edmonton, Alberta, T6G 2P5, Canada

Selim Kryczyński
Department of Plant Pathology
Warsaw Agricultural University
Nowoursynowska 166
02-766 Warsaw, Poland

Leslie C. Lane
Department of Plant Pathology
University of Nebraska
Lincoln, Nebraska 68583-0722, USA

N. Lukács
Institut für Physikalische Biologie
Heinrich-Heine Universität Düsseldorf
D-4000 Düsseldorf, Germany

Karl Maramorosch
Department of Entomology
Rutgers—The State University
New Brunswick, New Jersey 08903, USA

Alexander H. Purcell
Department of Entomological Sciences
University of California
Berkeley, California 94720, USA

Detlev Riesner
Institut für Physikalische Biologie
Heinrich-Heine Universität Düsseldorf
D-4000 Düsseldorf, Germany

W.G. van Slobbe
Denpasa
Cx-Postal 1351 CEP
66.000 Belém-PA, Brazil

James H. Tsai
Fort Lauderdale Research and Education Center
IFAS, University of Florida
3205 College Avenue
Fort Lauderdale, Florida 33314, USA

U. Wiese
Institut für Physikalische Biologie
Heinrich-Heine Universität Düsseldorf
D-4000 Düsseldorf, Germany

Chapter 1

A GENERAL METHOD FOR DETECTING PLANT VIRUSES

Leslie C. Lane

Leslie C. Lane is Associate Professor of Plant Pathology at the University of Nebraska, Lincoln, Nebraska. He received his BS in Chemistry in 1965 and PhD (Biochemistry) 1971 at the University of Wisconsin; 1971–1973 American Cancer Society. Post-doctorate, John Innes Institute, Norwich, England; 1975–1980 Assistant Professor. Associate Professor 1980 to date, University of Nebraska; American Phytopathological Society, American Society of Virology, Electrophoresis Society. Research interests: structure, genetics and classification of plant virus; gel electrophoretic and associated methods. Address: Department of Plant Pathology, University of Nebraska, Lincoln, NE 68583–0722.

CONTENTS

ABSTRACT

Plant viruses are reasonably uniform in size and stability. They generally sediment between 50 and 500S and are resistant to chelating agents. Few host components share these sedimentation and stability properties. Viruses can, therefore, be easily purified by ultracentrifugation. Subsequent gel electrophoresis of capsid proteins followed by silver staining detects and partially characterizes virsuses. Capsid protein size and yield often permit assignment of unknown viruses to a narrow range of taxa. Further analysis of partially purified virus usually permits specific diagnosis. Protein electrophoresis is a useful complement to dsRNA or viroid RNA electrophoresis for analyzing virus-like plant diseases of unknown etiology.

INTRODUCTION

In nature virus diagnosis has two extremes. At one extreme a specific disease may be recognizable by inspection. At the other extreme a causal agent may be totally mystifying. In the former case one confirms diagnosis by standard, specific diagnostic methods, such as serology or inoculating test plants, or, more recently, DNA-based methods. This I will term 'specific diagnosis'. In the latter case one must apply more general methods which supply clues about whether a virus is involved; and, if so, how specific diagnosis should proceed. This I will term 'general diagnosis'. Efficient general diagnosis leads to rapid progress in virology. For example, electron microscopy has led to identification of many potyviruses and gel electrophoresis has produced a similar taxonomic explosion among viroids. In both cases disease agents can be detected even where symptoms are erratic or nonexistent.

Four general methods are commonly applied to plant virus diagnosis. These are electron microscopy of particles in leaf extracts, electron microscopy of ultrathin tissue sections [1, 2], light microscopy of tissue strips [3], and gel electrophoretic analysis of double-stranded RNA [4]. Each has advantages and disadvantages. Low virus concentrations or unstable virions can frustrate direct electron microscopy. Icosahedral viruses can be difficult to distinguish from host components. Negative stains degrade some viruses. Thin sectioning is painstaking and time consuming. Furthermore, viruses invade tissues erratically; one can inadvertently sample inappropriate tissue. Light microscopy is still an art; plant virus inclusions are often indistinct. Tissues generally contain only traces of double-stranded RNAs and many viruses yield dsRNAs of similar size.

My laboratory has developed a fifth general diagnostic method which exploits the distinct biophysical properties of viruses. This method emphasizes the virus capsid protein which is usually the major virus-specific product in infected tissue. The method shares many advantages with gel electrophoretic diagnosis of viroids.

MATERIALS AND METHODS

General

Reagent purity is generally not critical. We normally use analytical reagent grade chemicals for handling purified virus. Chemicals can be obtained from standard supply houses or Sigma Chemical Co., St. Louis, Missouri.

Buffers

1. Extraction buffer—0.2 M diammonium citrate adjusted to pH 6.5 with ammonium hydroxide.
2. Neutral phosphate buffer—25 mM monobasic sodium phosphate plus 25 mM dibasic sodium phosphate.
3. Cracking buffer (sample buffer for sodium dodecyl sulfate (SDS) gel electrophoresis)—2% w/v SDS, 0.004% w/v crystal violet, 10% w/v Ficoll (sucrose is an adequate substitute), 2 mM disodium EDTA and 1/20 volume of 3 M Tris-HCl, pH 8.8 (36.3 g Tris, 3.7 ml reagent grade HCl, 69 ml distilled water). Add 1/50 volume of 0.5 M dithiothreitol immediately before use. 1–5% v/v (final concentration) of mercaptoethanol can substitute for dithiothreitol.

Minipurification [5]

1. Grind 2 g plant tissue in 13 ml extraction buffer with 0.15 ml, 0.25 M iodoacetamide and 0.15 ml, 1 M sodium diethyldithiocarbamate (DIECA) in a small Waring blender cup.
2. Express through wet muslin, centrifuge 10 min at 7K in Sorvall HS4 swinging bucket rotor.
3. Pour the supernatant into a polycarbonate centrifuge tube (for Beckman type 30 rotor), mix with 1 ml 33% Triton X-100, underlay with 5 ml 20% sucrose in extraction buffer.
4. Centrifuge for two hours at 27K in a Beckman type 30 rotor or 90 min at 35K in a Ti50.2 rotor, discard the supernatant; wash the sides of tube with distilled water.
5. Suspend the pellet in 1 ml 1/2 × concentrated extraction buffer.
6. Transfer to a 3 ml pyrex tube and centrifuge for 10 min at 7K.
7. Transfer the supernatant to a polycarbonate tube (for Ti50 rotor), add 3 ml of extraction buffer and underlay with 1.5 ml 20% sucrose in extraction buffer.
8. Centrifuge at 45K for 90 min in a Ti50 rotor.

9. Suspend pellets in neutral phosphate buffer. Use a smaller volume (typically 25–50 µl) for tiny pellets, or a larger volume (up to 1 ml) for large pellets.
10. Analyze the preparation by SDS gel electrophoresis, electron microscopy, or serology.

Phenol Extraction of Proteins

1. Bring the volume of the minipurified virus solution to about 0.5 ml, add 50 µl of 10% SDS, boil for 1 min, then extract by vortexing with 0.5 ml of water-saturated phenol. Keep phenol above $5°C$ to prevent freezing.
2. Centrifuge 5 min at 7K, withdraw the lower (phenol) phase leaving behind the interphase and upper phase. The upper phase can be saved to recover nucleic acids.
3. To the lower (phenol) phase, add 2.5 ml 0.1 M NH_4Ac in methanol, mix, store for two hours at $-20°C$ or one hour at $-80°C$ to precipitate protein.
4. Centrifuge 10 min at 7K in a swinging bucket rotor, discard the supernatant and wash the pellet with absolute methanol to remove residual phenol. Centrifuge again to compact the pellet.
5. Dry under vacuum, dissolve in 35 µl (or a larger volume if necessary) of cracking buffer.
6. Heat the sample 2 min at $100°C$ and electrophorese a 5–10 µl aliquot.
7. Nucleic acids can be precipitated from the aqueous phase and electrophoresed on agarose gels.

Gel Electrophoresis

Our methods are based on those of Studier [6]. Gels are 0.75 mm thick and are linear gradients of 8 to 25% polyacrylamide. Compositions are 8% T, 0.25% C and 25% T, 1.5% C. The stacking gel contains 3% T, 30% C with piperazine diacrylamide rather than methylene bisacrylamide as the cross-linker. Running buffer is pH 8.5 and is prepared by volumetric addition of concentrated HCl rather than by adjusting with the pH meter. Samples in cracking buffer are heated 2 min at $100°C$ immediately before electrophoresis. Aliquots of 5–10 µl are applied to gels. Loading is determined by trial and error. For dilute viruses we mix the sample with one-third volume of cracking buffer and apply the maximum volume. For concentrated viruses we dilute the virus solution many fold to prevent overloading. We electrophorese overnight at about 100 V (6 V/cm). The tracking dye should be close to the bottom of the gel at the end of the run. Specific details of electrophoresis and associated procedures are available from the author.

Staining

We employ the silver development method of Morrissey [7]. Since our gels are half the thickness of his we have reduced the soaking times two-fold.

RESULTS AND DISCUSSION

Purification

The only major proteinaceous plant components which sediment with viruses are ribosomes and membrane fragments. Chelating agents (such as citrate) disrupt ribosomes and non-ionic detergents disrupt membrane fragments. Sedimenting the virus through a sucrose cushion prevents host proteins such as rubisco (20S) from reaching the pellet.

The ammonium citrate buffer extracts viruses efficiently. We have yet to find a buffer which more efficiently extracts any virus. The pH and ionic strength are close to physiological and should stabilize viruses. The relatively high ionic strength minimizes polyanion-polycation interactions which could precipitate viruses. Iodoacetamide prevents disulfide oxidation and inactivates sulfhydryl proteases which are the major proteases of crude plant extracts. DIECA, which chelates zinc, prevents some capsid proteins, notably cucumber mosaic virus, from aggregating. The aggregation probably involves enzymatic oxidation. Iodoacetamide and DIECA are not essential for most viruses. Membrane dissociation requires an excess of non-ionic detergent. Chlorophyll (green) in the pellet following ultracentrifugation indicates insufficient detergent. The volume of the sucrose cushion and the ultracentrifugation time can be adjusted to maximize recovery of virus or to minimize host contaminants.

Many plant tissues require a second cycle of high- and low-speed centrifugation to give proteins of adequate quality for SDS gel electrophoresis. A third cycle is occasionally useful. Tissues which are rich in gums or carbohydrates (banana leaves, for example) yield very large pellets which contain little protein. In these cases, phenol extraction of protein concentrates protein sufficiently to give bands detectable on gel electrophoresis. Tissues rich in tannins and phenolics may require special handling. Techniques that should be considered are grinding with liquid nitrogen, adsorbing precipitants with excess low molecular weight protein, and adding other antioxidants.

Gel Electrophoresis

Individual labs will normally adapt their own SDS gel electrophoresis protocols for diagnosis. Procedures, however, cannot be adapted blindly. Buffer preparation, detergent levels, gel pore size, and protein reduction must be carefully considered. Salt concentration of electrophoresis buffers is more important than exact pH. It is not permissible to back titrate if one overshoots a pH. Comparing solutions with conductivity meters is usually more helpful than comparing with pH meters.

Virology literature expresses detergent concentration in absolute terms. In fact, proteins bind roughly twice their weight and membranes bind roughly 10 times their weight of SDS. SDS: protein ratio and SDS: lipid ratio are generally more instructive than absolute SDS concentration. Sample buffers must contain enough SDS to saturate protein and lipid and enough left over to form micelles (approximately 0.1% for moderate buffer concentrations). Once protein has been denatured, one need only be sure that free SDS concentration does not fall below 0.1%. Fortunately, one can easily test for adequate detergent levels. Simply repeat the experiment with a higher level of SDS. If results improve, the earlier level was inadequate.

For electrophoresis with Laemmli buffers, chloride concentration of running gel buffer is critical. Aggregation (micellization) of SDS depends on salt concentration. If chloride concentration in the running gel drops below a critical level, SDS will no longer be aggregated and electrophoretic resolution will degenerate. The running buffer pH suggested by Laemmli [8] (8.8) is perilously close to the minimum required to maintain micelles at 0.1% SDS. We avoid this problem by using a lower pH, 8.5 (i.e. adding more HCl). Adding a fixed volume of HCl with a pipette gives a more reproducible buffer than adjusting pH with a meter.

Occasionally electrophoresis apparati can be a problem. Buffer wells must be large enough to maintain pH during the run. Checking pH of upper and lower wells after a run guards against inadequate buffer capacity. Permissible variation from the initial pH is determined only by experience. Fortunately the buffering capacity of Laemmli buffers is high and buffer titration is less of a problem than for most systems.

Sulfhydryl groups are notoriously sensitive to oxidation. Either incomplete reduction of proteins or partial reoxidation after reduction can lead to broad bands on SDS gel electrophoresis. Dilute solutions of reducing agents or reduced proteins oxidize within a few days. Freezing solutions does not prevent reoxidation. Cooling solutions increases oxygen solubility and freezing creates cracks which admit oxygen. Solutions of thiols should be freshly prepared. For storage, reduced proteins should be alkylated [9] to block thiol groups. Oxidation problems are detected by either using fresh reducing agents or comparative electrophoresis of reduced proteins and reduced, alkylated proteins. For extended electrophoresis, incorporating an anionic thiol (e.g. 0.1% v/v mercaptopropionic acid) into the upper buffer well inhibits protein reoxidation [9].

The Morrissey method stains gels evenly and reproducibly. We have varied the concentration of individual reagents by factors of two and still get satisfactory staining. Silver staining protocols, however, must be adopted cautiously. Literature procedures are designed for gels of specific thickness and specific pore size. Thinner gels require shorter soaking times. Smaller pore sizes require longer soaking times. Solution volumes should remain in constant ratio to gel volume. Soaking times are easily optimized by varying them systematically with different slices from a single gel. Increasing gel thickness beyond 0.75 mm increases staining time without compensating benefits.

Before attempting virus diagnosis by protein gel electrophoresis, one should develop confidence in protein gel electrophoresis. This means generating interpretable patterns for standard proteins. One should diagnose well-characterized, high-level viruses before attempting diagnosis of obscure agents.

General Diagnosis

Protein bands which are unique to diseased tissue indicate virus infection (see Table 1.1). Virus infection can be confirmed by showing that the extra protein is reproducible and ultimately by specific diagnosis. The apparent size of a protein and its relative amount are clues to virus identity. Comparison of protein size and band intensity to protein sizes and virus yields of major virus taxa narrows (often considerably) the potential identity of the virus. Obviously, parallel healthy controls are often critical to recognizing virus-specific bands. Where virus-specific bands are faint, knowledge of the range of variation of healthy controls can minimize overoptimistic interpretations. Where the origin of a particular band is in question, it may help to purify virus from tissue of several different hosts.

Protein patterns from partly purified preparations can be complex. Such preparations normally contain recognizable host proteins. The large and small subunits of rubisco (about 55 kd and 12 kd) are generally prominent. They presumably arise from rubisco aggregates or complexes. Preparations usually contain several larger host-specific proteins (roughly 100 kd) which can serve as markers for host identification. Individual plant species have characteristic protein contaminants. For example, soybeans are generally rich in rubisco and contain a 29 kd protein. *Chenopodium quinoa* and *Gomphrena globosa* are relatively rich in host protein contaminants. Solanaceous plants generally have few major contaminants. Protein patterns are surprisingly insensitive to age or physiological condition of the host. There are occasional 'erratic' protein contaminants. These are often protein families such as ribosomal proteins or histones. Where ultracentrifuge pellets are rich in gums, these must be removed (by phenol extraction) to concentrate proteins sufficiently to yield detectable bands.

Gel electrophoretic patterns often contain additional clues to virus identity. For example, subgenomic RNAs migrate into protein gels and stain with silver. Potyviruses are often contaminated with traces of characteristic inclusion body proteins. Partial degradation of virus capsid protein may yield characteristic fragments. Silver development imparts characteristic colors to protein bands. One must also remember that multiple infections and satellite viruses can add complexity to a protein pattern.

Specific Diagnosis

Once general diagnosis has confirmed a virus and narrowed the range of virus taxa, one must guess which particular virus is involved. Several information sources can improve the guess. First of all, local experience is invaluable. In the absence of such experience, specific crop disease manuals or *Review of Plant Pathology* should be consulted. The most difficult cases are those in which the virus is previously unidentified. In such cases one needs to narrow the range of taxa by further physical characterization. In ideal circumstances the virus will betray its affinity by cross-reacting with antiserum to a known virus. In some cases one will simply have to report virus properties and imply relationships or lack thereof.

Since protein electrophoresis requires only part of a purified sample, the remainder can be examined by electron microscopy, serology, or other techniques. The initial host as well as virus morphology, capsid protein size, and yield provide clues as to which techniques and which antisera should be tested. Immunodiffusion is convenient for icosahedral viruses and electron microscope serology or western blots are convenient for elongated viruses. For spherical viruses, phenol extraction and gel electrophoresis of viral RNA can be useful.

Host range, though time consuming, is also useful for specific identification. Partial purification can facilitate mechanical transmission especially for stable viruses. However, one must be very careful to prevent contamination during purification. For example, if one purifies

Table 1.1 Plant virus groups

Class	Examples	Approx. number of viruses	Length (or dia.)	Sed. Coeff.	Nucleic acid size	Vector	Protein size	Host range	Yield
Rod-shaped viruses									
TOBAMOVIRUSES	tobacco mosaic ribgrass mosaic	10	300 nm	195S	ssRNA 6.4 kb	?	17K	Broad	High
FUROVIRUSES (bipartite)	soil-borne wheat mosaic	10	300 nm, 150 nm	172, 211S	ssRNA 7.2 kb, 3.4 kb	Fungi	20K	Narrow	Med.
TOBRAVIRUSES (bipartite)	tobacco rattle pepper ringspot pea early browning	3	200 nm, 50–120 nm	300, 155–243S	ssRNA 7.2 kb 1.9–4.1 kb	Nematodes	20K	Broad	Low to med.
HORDEIVIRUSES (tripartite)	barley stripe mosaic pea semilatent lychnis ringspot	3	110–160 nm	182, 197,201S	ssRNA 3.9 kb, 3.3 kb 2.8 kb	Seed-borne	22K	Narrow	High
Filamentous viruses									
POTEXVIRUSES	potato virus X white clover mosaic	25	480–580 nm	110–125S	ssRNA 6.6 kb	Fungi	24–27K	Moderate	High
CARLAVIRUSES	carnation latent potato virus S	10	620–690 nm	156–172S	ssRNA 7.2 kb	Aphids (nonpersistent)	27–33K	Narrow	Low
POTYVIRUSES	potato virus Y bean common mosaic maize dwarf mosaic	100	680–900 nm	140–155S	ssRNA 10.8 kb	Aphids (nonpersistent)	35K	Narrow	Low to med.
WHEAT STREAK MOSAIC VIRUS	ryegrass mosaic oat necrotic mottle	5	700 nm	165S	ssRNA 8.8 kb	Mites	45K	Narrow	Med.
SWEET POTATO MILD MOTTLE VIRUS		1	800-950 nm	155S	ssRNA ?	Whiteflies	38K	Broad	?
TENUIVIRUSES	rice stripe maize stripe rice hoja blanca	5	very thin filaments	50–70S	ssRNA 5.9 kb, 4.4 kb 3.1 kb, 2.8 kb	Planthoppers[1]	32K	Narrow to moderate	Low?

WHEAT SPINDLE STREAK MOSAIC VIRUS	5	barley yellow mosaic, oat mosaic	300–2000 nm highly flexuous	?	ssRNA ?	Fungi	36K	Narrow	Very low
CLOSTEROVIRUSES	10	sugarbeet yellows, citrus tristeza, apple chlorotic leafspot	600–2000 nm highly flexuous	95–140S	ssRNA 7.2–13.4 kb	Aphids (nonpersistent)	22K	Moderate	Low
CAPILLOVIRUSES	12	apple stem grooving, lilac chlorotic leafspot	640 nm	100S	ssRNA 7.8 kb	Unknown	27K	Moderate	?
Spherical viruses, DNA-containing									
CAULIMOVIRUSES	10	cauliflower mosaic, dahlia mosaic	50 nm	220S	dsDNA 4 kb	Aphids (semipersistent)	Several proteins	Narrow	Med
GEMINIVIRUSES (bipartite)	20	cassava latent, bean golden mosaic	20 nm 'twin particles'	75S	ssDNA 2.7 kb(2)	Whiteflies[1]	28–31K	Narrow	Low
(monopartite)		maize streak, beet curly top			2.7 kb(1)	Leafhoppers[1]	27–31K	Broad (BCTV)	
Spherical viruses, dsRNA-containing									
PHYTOREOVIRUSES	3	rice dwarf, wound tumor	65–70 nm	500S	dsRNA 12 components	Leafhoppers[1]	Several proteins	Narrow	Low?
FIJIVIRUSES	10	fiji disease, maize rough dwarf	65–70 nm	500S	dsRNA 10 components	Planthoppers[1]	Several proteins	Narrow	Low?
CRYPTIC VIRUSES (virus-like particles)	20	beet cryptic, carnation cryptic	30 nm	120S	dsRNA multiple 0.5–1.5 kb	Seedborne[1]	54K	Narrow	Very low
Spherical viruses, ssRNA-containing, monopartite									
MAIZE CHLOROTIC DWARF VIRUS	2	rice tungro	30 nm	183S	ssRNA 10 kb	Leafhoppers[1]	Three proteins	Narrow	Low?
MARAFIVIRUSES	2	maize rayado fino, oat blue dwarf	25 nm	120S	ssRNA 6.6 kb	Leafhoppers[1]	21K	MRFV narrow OBDV-broad	Low?
TYMOVIRUSES	20	turnip yellow mosaic, eggplant mosaic	25 nm	55, 115S	ssRNA 6.3 kb	Beetles	20K	Narrow	High
LUTEOVIRUSES	15	barley yellow dwarf, beet western yellows	25 nm	115–130S	ssRNA 6.3 kb	Aphids[1] (persistent)	25K	Narrow	Very low

[1]Not mechanically transmissible
[2]Some bottom component particles contain two middle component RNAs
[3]Subgenomic RNA

Table 1.1 Plant virus groups (contd.)

Class	Examples	Approx. number of viruses	Length (or dia.)	Sed. Coeff.	Nucleic acid size	Vector	Protein size	Host range	Yield
TOMBUSVIRUSES	tomato bushy stunt cymbidium ringspot	5	25 nm	130–140S	ssRNA 4.7 kb	Fungi(?)	41K 85K (minor)	Wide	High
CARMOVIRUSES	carnation mottle turnip crinkle	10	25 nm	125S	ssRNA 4.7 kb	(?)	40K	Broad	High
SOBEMOVIRUSES	southern bean mosaic sowbane mosaic	5	25 nm	115S	ssRNA 4.7 kb	Beetles	30K	Narrow	High
PHLEUM MOTTLE VIRUS	panicum mosaic cocksfoot mild mosaic	8	25 nm	105–115S	ssRNA 4.7 kb	Beetles	25K	Narrow	High
NECROVIRUSES	tobacco necrosis cucumber necrosis	2	25 nm	120S	ssRNA 4.7 kb	Fungi	23K	Wide	Low
MAIZE WHITE LINE MOSAIC VIRUS		1	35 nm	120S	ssRNA 3.9 kb	Soilborne[1]	32K	Narrow	High
PARSNIP YELLOW FLECK VIRUS	dandelion yellow mosaic	3	50–70 nm (membrane)	?	ssRNA 4.8 kb	Aphids helper dependent	31,26,24K	Broad	?
Spherical viruses, ssRNA-containing, bipartite									
DIANTHOVIRUSES	carnation ringspot red clover necrotic mosaic	6	34 nm	135S	ssRNA 4.7 kb, 1.6 kb	Nematodes	38K	Broad	High
PEA ENATION MOSAIC		1	28 nm	90, 110S	ssRNA 5.3 kb, 4.5 kb	Aphids (persistent)	22K	Narrow to moderate	Low to mod.
COMOVIRUSES	cowpea mosaic bean pod mottle	15	25 nm	60, 100, 125S	ssRNA 6.9 kb, 3.4 kb	Beetles	42K 22K	Narrow	High
FABAVIRUSES	broad bean wilt lamium mild mosaic	2	25 nm	63,100,126S	ssRNA 6.9 kb, 3.4 kb	Aphids (nonpersistent)	42K 22K	Broad	Med.
NEPOVIRUSES (tobacco ringspot subgroup)[2]	grapevine fanleaf	10	25 nm	50, 100, 125S	ssRNA 7.8 kb, 4.4 kb	Nematodes	55K	Wide	High

(tomato ringspot subgroup)	chicory yellow mottle	3	25 nm	50, 120, 125S	ssRNA 7.8 kb, 7.0 kb				
(cherry leafroll subgroup)	peach rosette mosaic	6	25 nm	50, 110, 125S	ssRNA 7.8 kb, 6.0 kb				
(tomato black ring subgroup)	tomato top necrosis	6	25 nm	50, 100, 125S	ssRNA 7.7 kb, 5.0 kb				
Spherical viruses, ssRNA-containing, tripartite									
CUCUMOVIRUSES (tripartite)	cucumber mosaic peanut stunt tomato aspermy	3	25 nm	100S	ssRNA 3.4 kb, 3.1 kb 2.2 kb, 0.8 kb[3]	Aphids (nonpersistent)	25K	Broad	Low to high
BROMOVIRUSES	brome mosaic broad bean mottle cowpea chlorotic mottle	6	25 nm	85S	ssRNA 3.4 kb, 3.1 kb 2.2 kb, 0.9 kb[3]	Beetles	20K	Narrow	High
ILARVIRUSES	tobacco streak apple mosaic prunus necrotic ringspot	10	25–35 nm irregular	80–120S	ssRNA 3.4 kb, 3.1 kb 2.2 kb, 0.9 kb[3]	(?)	25K	Broad	Low to high
Miscellaneous viruses									
ALFALFA MOSAIC VIRUS (tripartite)	(bacilliform)	1	20 nm width variable lengths	50–100S	ssRNA 3.4 kb, 3.1 kb 2.2 kb, 0.9 kb[3]	Aphids (nonpersistent)	24K	Wide	High
TOMATO SPOTTED WILT (ambisense RNA)	similar to animal bunyaviruses	1	80 nm pleomorphic (membrane)	530S	ssRNA 8.1 kb, 5.9 kb 5.3 kb, 4.1 kb	Thrips	Several proteins	Wide	High
BADNAVIRUSES	cocoa swollen shoot rubus yellow net rice tungro bacilliform	8	28 × 125 nm		dsDNA	Mealybugs,[1] aphids,[1] leafhoppers[1]	?	Narrow	Low?
RHABDOVIRUSES (-RNA)	potato yellow dwarf maize mosaic broccoli necrotic yellows	20	bullet-shaped with membrane 70 × 200 nm	900S	ssRNA 14.1 kb usually[1]	Leafhopper, aphids	Several proteins	Narrow	Med?

[1]Not mechanically transmissible
[2]Some bottom component particles contain two middle component RNAs
[3]Subgenomic RNA

TMV in parallel with other viruses, all samples will generally contain TMV. Contamination of sensitive specimens is avoided by keeping them far away from TMV and by inoculating hosts that are resistant to TMV. Those who are not fluent with both physical and microbiological concepts of purity are likely to experience 'baffling' contaminants. These invariably turn out to be viruses which are below physical detection limits, but still infectious.

Where different virus isolates give protein bands of similar size and intensity, additional techniques can confirm identity or establish differences. Peptide mapping is especially convenient. We have developed simple chemical methods (cyanogen bromide and formic acid cleavage) which give reproducible and characteristic fragment patterns [10]. For icosahedral viruses, gel electrophoresis of whole virions is convenient. These provide additional information on relationships among virus isolates.

Other Applications

The partial purification technique can also be used to optimize large-scale purification. One can use different buffers. More acidic solutions (particularly pH about 4.5) give cleaner virus preparations, but many viruses are unstable at low pH. Non-ionic detergents are generally interchangeable for virus purification; however, other detergent classes can be used. Sodium dodecyl sulfate gives very clean preparations but only for viruses which tolerate it. Centrifugation times can be varied to minimize impurities or maximize virus yield. Virus extracts can be heated to various temperatures prior to centrifugation. In our hands this rarely improves purification except for the most stable viruses. For difficult to purify viruses, various hosts, tissues, and growth temperatures can be tested in parallel to optimize yield and purity. Inactivation of viruses by host components can be detected by mixing purified virus with plant tissue prior to purification. If recovery is poor, procedures can be modified to improve recovery.

Protease treatment of crude virus preparations can be useful. Viruses in general are resistant to proteases and many host proteins are susceptible. Unfortunately, major host proteins such as rubisco tend to resist proteases. Some capsids contain protein 'tails' which are cleaved by proteases. In these cases protease treatment reduces the apparent size of the capsid protein. Thermolysin is especially useful because it is stable and highly active, but can be readily inactivated by chelating agents.

For pathologists who specialize in a specific host, it may be helpful to tailor the purification to minimize contaminants from the specific host.

Different buffers, different centrifugation times, or additional clarification schemes might be considered. Fulton [11] suggested that antisera to host contaminants might be especially useful for clarifying plant extracts. His suggestion has been ignored, probably for fear that antiserum will contaminate viruses. If fact, antiserum treatment is likely to replace impurities that are difficult to remove (particularly carbohydrates) with easily removable contaminants. Furthermore, antiserum contamination would be 'invisible' for purposes of preparing antiserum to viruses. Though phenol extraction also removes carbohydrates from virus preparations, it denatures virus precluding subsequent analysis by techniques which require virus integrity.

For viruses that produce stable inclusions, analogous partial purification procedures using low-speed centrifugations permit virus diagnosis by electrophoresis of inclusion proteins. Potyvirus infections can give higher yields of inclusion proteins than of virion proteins.

Overview

As with all diagnostic techniques, there are limitations. The method requires an ultracentrifuge. It can handle a limited number of samples (roughly 24 per day). While this is normally ample for general diagnosis it is clearly inadequate for large-scale specific diagnosis. Tissues that are rich in carbohydrates or oxidizing agents can pose specific problems. Unstable viruses, viruses within inclusions, and viruses at low levels (phloem limited) pose additional problems. Instability is not as widespread as might be expected. For example, nucleocapsids of membranous viruses such as rhabdoviruses and tomato spotted wilt can be purified and detected by these methods.

The technique is quick, handles many samples, and is surprisingly general. Since it can handle tissue directly from the field, viruses can be diagnosed before exposure to greenhouse contaminants. Multiple infections can be recognized if they arise from viruses with distinct capsid proteins. The technique gives taxonomically useful information, i.e., information that narrows the range of potential taxa facilitating specific diagnosis.

We have diagnosed many viruses, including alfalfa mosaic virus, bromoviruses, carlaviruses, comoviruses, cucumoviruses, nepoviruses, potexviruses, potyviruses, rhabdoviruses, sobemoviruses, tobamoviruses, tobraviruses, tomato spotted wilt virus, tymoviruses, and wheat streak mosaic virus. We also use the technique to monitor purity of virus cultures in the greenhouse. Versatility and information richness make the technique a delight for those who enjoy problem solving, but the potential for overly concrete interpretation makes it a potential nightmare for those who prefer diagnosis by recipe.

SPECIFIC EXAMPLES

Hollyhock Mosaic

Leaves from hollyhocks (malvaceae) with interveinal yellowing yield large amounts of a virus with a 20 kd capsid protein. Substituting SDS for Triton X-100 gives similar virus yields. Yield, protein size, and virion stability suggest a tymovirus. The virus infects only hollyhocks. Greenhouse symptoms are mild and often difficult to recognize.

The virus reacts weakly with antiserum to turnip yellow mosaic virus in double diffusion. On electrophoresis at pH 7 it migrates anionically showing two distinct components with about half the mobility of turnip yellow mosaic virus. The faster migrating component coincides with the ultracentrifugal bottom component and the slower component corresponds with the ultracentrifugal top component (empty shells).

Hollyhock mosaic is a tymovirus. Host range and electrophoretic mobility are distinct from those of okra mosaic, the only other tymovirus which naturally infects malvaceous hosts. Hollyhock weevils, which are of European origin, are common in Nebraska and are plausible vectors.

Viruses of *Mertensia virginica*

Virginia bluebell, *Mertensia virginica* (Boraginaceae), is a common, early spring garden flower in Lincoln. It often expresses severe mosaic symptoms. Infected plants yield preparations with 17 kd protein, 29 kd protein, or both. The 17 kd protein is slightly smaller than TMV protein and stains a different color with silver. Preparations with 17 kd protein contain TMV-like rods. Cyanogen bromide, which cleaves after methionine, produces discrete fragments from the 17 kd protein, but does not digest TMV-type strain protein. The high methionine content of the 17 kd protein suggests ribgrass mosaic virus. Direct comparison to known ribgrass mosaic virus confirmed the identity. Subsequently, ribgrass mosaic virus was isolated form both *Plantago major* (Plantaginaceae) and *Hesperis matronalis* (Cruciferae). Symptoms in the former were mild and difficult to detect. Cucumber mosaic is the only virus disease of *Mertensia* reported in *Review of Plant Pathology*. Indeed, preparations containing the 29 kd protein reacted in immunodiffusion with cucumber mosaic virus antiserum.

Cherry Rasp Leaf Virus

Partial purifications from several weeds with virus-like symptoms gave a triplet of bands of apparently equal intensity, ranging in size from 20 kd to 24 kd. The three bands stain different colors with silver. Partial purification in the presence of SDS gives exceptionally clean virus preparations. Fixed virus is icosahedral by electron microscopy. The virus infects *Chenopodium quinoa, Datura stramonium,* and *Nicotiana benthamiana* as well as many other hosts. In most hosts symptoms are mild to invisible. By late spring, the virus becomes impossible to recover or transmit in the greenhouse and can be maintained only in a cooled chamber. The virus contains two RNAs similar in size to tobacco ringspot virus RNAs. The virus gives a reaction of identity in double diffusion with a Scottish strain of cherry rasp leaf virus [12]. The proteins of the two strains, though similar, differ slightly in apparent size and the two strains are distinct by virion gel electrophoresis.

Cherry rasp leaf virus has been recovered from many weed species and is common in Nebraska. This virus would have been difficult to identify by standard techniques because of its low levels in tissue, mild symptomatology, and aversion to high temperatures. Its distinctive protein pattern facilitated identification and comparison to a known strain.

REFERENCES

1. Edwardson, J.R., and R.G. Christie. Use of virus-induced inclusions in classification and diagnosis. *Ann. Rev. Phytopathol.*, 16, 31, 1978.
2. Francki, R.I.B., R.G. Milne and T. Hatta. *Atlas of Plant Viruses*, 2 vols. Boca Raton (Florida, USA): CRC Press, 1985.
3. Christie, R.G., and J.R. Edwardson. Light microscopic techniques for detection of plant virus inclusions. *Plant Dis.*, 70, 273, 1986.
4. Valverde, R.A., J.A. Dodds and J.A. Heick. Double-stranded ribonucleic acid from plants infected with viruses having elongated particles and undivided genomes. *Phytopathology*, 76, 459, 1986.
5. Lane, L.C. Propagation and purification of RNA plant viruses. *Meth. in Enzymol.* 118, 687, 1986.
6. Studier, F.W. Analysis of bacteriophage T7 early RNAs and proteins on slab gels. *J. Mol. Biol.*, 79, 237, 1971.
7. Morrissey, J.H. Silver stain for proteins in polyacrylamide gels: a modified procedure with enhanced uniform sensitivity. *Anal. Biochem.*, 117, 307, 1981.
8. Laemmli, U.K. Cleavage of structural proteins during the assembly of the head of bacteriophage T4. *Nature*, 227, 680, 1970.
9. Lane, L.C. A simple method for stabilizing protein-sulfhydryl groups during SDS-gel electrophoresis. *Anal. Biochem.*, 86, 655, 1978.
10. Skopp, R.N., and L.C. Lane. Fingerprinting of proteins cleaved by cyanogen bromide. *Appl. and Theoret. Electrophoresis*, 1, 61, 1989.
11. Fulton, R.W. Purification and some properties of tobacco streak and Tulare apple mosaic viruses. *Virology*, 32, 153, 1967.
12. Jones, A.T., M.A. Mayo and S.J. Henderson. Biological and biochemical properties of an isolate of cherry rasp leaf virus from red raspberry. *Ann. Appl. Biol.*, 106, 101, 1985.

Chapter 2

FATAL YELLOWING OF OIL PALMS: SEARCH FOR VIROIDS AND DOUBLE-STRANDED RNA

E. Beuther, U. Wiese, N. Lukács, W.G. van Slobbe, and
D. Riesner*

*Detlev Riesner is Full Professor of Physical Biology at Heinrich-Heine-Universität Düsseldorf, FRG. He received his master in physics 1966 at Hannover Institute of Technology and PhD (Biophysical Chemistry) 1970 at Braunschweig Institute of Technology. Post-doctoral fellow of Princeton University, New Jersey, USA, 1973. Lecturer in Biophysical Chemistry and Molecular Biology at Hannover Medical School, 1977.
1977–1980 Associate Professor of Biochemistry at Darmstadt Institute of Technology. 1980 to date Düsseldorf. 1985 Guest Professor at University of California, San Francisco. Research interests: physical chemistry of nucleic acids, structure-function of viroids, molecular biology of forest decline, structure of the scrapie causative agent. Address: Institut für Physikalische Biologie, Heinrich-Heine-Universität Düsseldorf, Universitätsstr. 1, D-4000 Düsseldorf 1, FRG.

CONTENTS

ABSTRACT

Fatal yellowing is a serious disease of still unknown origin affecting oil palms in several regions of Central and South America. In the present study a search for viroids and viroid-like RNAs in oil palms was performed. For such a search standard methods like the two-dimensional gel electrophoresis and the return gel electrophoresis were applied to nucleic acid extracts. Although RNAs with viroid-like gel electrophoretic behavior were detected, the presence of the known viroids was excluded by hybridization tests with probes for potato spindle tuber viroid (PSTVd), coconut cadang-cadang viroid (CCCVd), or *Coleus blumei* viroid (CbVd1). In order to detect double-stranded RNA (dsRNA), antibodies against dsRNA were used, and the viroid-like RNAs could be identified as dsRNA. The same pattern of dsRNA was found in healthy as well as in diseased oil palms; thus, these dsRNAs are most probably not of viral origin and cannot be part of the causative agent of fatal yellowing. Other possibilities for the origin of the dsRNA are being considered. A method to avoid misinterpretation by the return gel electrophoresis is presented, i.e. to differentiate unequivocally between circular and dsRNA.

INTRODUCTION

Fatal yellowing is a serious disease affecting oil palms (*Elais guineensis* Jacq.) in several regions of Central and South America for more than 20 years [1]. Fatal yellowing [2] is also known as lethal spear rot, pudrición del cagollo, and amar elecimento fatal [3]. The first symptom is a chlorosis of the youngest five to six fronds, which are closest to the centre of the crown. In early stages of the disease, patches of chlorotic or even necrotic tissue of the spear leaf are affected by a wet rot. One or more spear leaves break near their bases. The chlorosis may be detected simply because the fronds in the lower part of the tree are dark green. Normally it takes 10 to 24 months from the appearance of the first symptoms to the final dying back of the palm.

During recent years the disease spread very rapidly resulting in severe economical losses (P. Kastelein, personal communication). Its epidemical character strongly suggested that it is an infectious disease, but in spite of intensive research [2–6] neither the causative agent nor a vector of the pathogen could be identified until now. Recently, viroid-like RNAs were detected by return gel electrophoresis [3]. They were found in leaves and roots of older oil palms with and without symptoms of fatal yellowing, but not in symptomless younger trees. Indeed, several diseases of palms are described to be caused by viroids and viroid-like pathogens [7,8]. Thus it was of particular interest to follow the early findings of Singh and colleagues and to clarify whether viroids are involved in fatal yellowing.

In the present study, the occurrence of viroids in oil palms grown at DENPASA (Dendé do Para, South America, Brazil) was investigated by two-dimensional gel electrophoresis, by return gel electrophoresis, and by hybridization tests. It could be shown by physico-chemical and immunobiological techniques that the 'viroid-like' RNAs are in fact double-stranded RNAs. At present this does not imply, however, that the dsRNA is part of a virus.

MATERIALS AND METHODS

Samples

Samples were taken at DENPASA oil palm estate near Belém in Brazil. At this location the first observations of fatal yellowing were made in 1974. For several years the progression of the disease was very slow, but from 1983 it became very rapid, resulting until 1990 in the loss of more than 15% of the 720,000 trees planted at DENPASA. For our analysis leaves were taken from palms of different age. In each age group palms without symptoms, with early symptoms and with advanced symptoms of fatal yellowing were selected. Up to four leaves were cut from each palm.

Chemicals

Acrylamide and bisacrylamide were purchased from Serva (Heidelberg, FRG), enzymes from Boehringer (Mannheim, FRG), alkaline phosphatase conjugated goat-anti-mouse IgG from Dianova (Hamburg, FRG). All other chemicals and solvents were of reagent grade from commercial sources.

Buffers

PBS-buffer: 10 mM KH_2PO_4/K_2HPO_4, 150 mM NaCl, pH 7.2.
NP-buffer: 50 mM NaH_2PO_4/Na_2HPO_4, pH 7.2.
STE-buffer: 50 mM Tris-HCl, 100 mM NaCl, 1 mM EDTA, pH 7.0
TBE-buffer: 89 mM Tris, 89 mM boric acid, 2.5 mM EDTA, pH 8.3.
TNE-buffer: 100 mM Tris-HCl, 100 mM NaCl, 10 mM EDTA, pH 8.0.
TE-buffer: 10 mM Tris-HCl, 1 mM EDTA, pH 7.5.
AP-buffer: 100 mM Tris-HCl, 100 mM NaCl, 50 mM $MgCl_2$, pH 9.5.
TK-buffer: 40 mM Tris-HCl, 100 mM NaCl, 10 mM Dithiothreitol
                6 mM $MgCl_2$, 2 mM spermidine, pH 8.0.

Extraction of Nucleic Acids

Procedure (a)
Nucleic acids were extracted from the oil palm leaves using the

procedure of Randles [9], with the modification of Singh et al. [3]. The column chromatography with CF11-cellulose for the elimination of colored material was replaced by a batch procedure: Nucleic acids were re-dissolved in STE-buffer containing 35% ethanol and added to 2.5g CF11-cellulose powder equilibrated in the same buffer. After gentle stirring for one hour the fluid was removed by slow centrifugation (600 xg), and the CF11-material was washed three to four times for 15 min with STE-buffer/35% ethanol. Bound nucleic acids were eluted with STE without ethanol by stirring the CF11 material two times for 30 min. The eluted nucleic acids were precipitated with 2.5 volumes of ethanol, pelleted by centrifugation, and dissolved in NP-buffer. The yield of purified nucleic acids was about 1 μg per gram of leaf material.

Procedure (b)

A markedly higher amount of nucleic acids—about 150 μg per gram of leaf—could be extracted by the following method. After addition of Al_2O_3-powder, 20 g of oil palm leaflets were pulverized under liquid nitrogen using mortar and pestle. The resulting powder was suspended in a solution containing 60 ml TNE-buffer, 60 ml Tris-buffer saturated phenol pH 8.0, 3 g sodium dodecyl sulfate (SDS), 0.6 g Na_2SO_3, and 0.12 g dithiothreitol. A homogeneous mixture was formed by shaking, and 60 ml chloroform was added. After centrifugation at 6600 xg, the aqueous phase was extracted once again with phenol/chloroform and subsequently twice with chloroform. The nucleic acid was precipitated by 2.5 volumes of 96% ethanol overnight, pelleted, and air-dried. The pellet was dissolved in 12 ml 2 × TE-buffer, and the resulting yellow-brownish solution was purified by a modification of the CF11-cellulose chromatography batch procedure described above. An aliquot of 2.4 ml was adjusted to STE-buffer/35% ethanol and added to one gram of CF11-cellulose powder equilibrated in the same buffer. After gentle mixing for one hour the fluid was removed by gravity sedimentation, and the CF11-material was washed three times with 30 ml STE-buffer/35% ethanol. The cellulose pellet was mixed with 4 ml STE-buffer, poured into a 10 ml plastic syringe which was plugged with siliconized glass wool. The eluate and 12 ml STE-buffer from a second elution were collected and precipitated with ethanol. Nucleic acids were recovered by centrifugation and dissolved in TE-buffer.

One-dimensional Gel Electrophoresis

Gel electrophoresis was carried out in 5% acrylamide (w/v), 0.16% bisacrylamide (w/v), 0.3% TEMED (N,N,N´,N´-tetramethyl-ethylendiamine) (v/v) in TBE-buffer, and 0.07% ammoniumperoxidisulfate (w/v) for starting the polymerization. A slab gel instrument BioRad model 220 (14.0 ×10.0 × 0.15 cm gel size) was used (BioRad Munich, FRG). Samples of 50 μg nucleic acids were mixed with an equal volume of the dye markers bromphenol blue (0.01%) and xylene-cyanol (0.01%) in glycerol-buffer. Electrophoresis was performed at 20°C and 220 V and was stopped for hybridization tests just before bromphenol blue left the gel, and for immunoblotting, before xylene-cyanol left the gel.

Two-dimensional Gel Electrophoresis

Two-dimensional gel electrophoresis was carried out according to Schumacher et al. [10] with the modifications of Beuther et al. [11] Five per cent resp. 7.5% acrylamide gels in TBE with 3 M urea in addition were used. Samples were run in the first dimension under denaturing conditions (60°C, 1 hr, 500 V) from top to bottom and in the second dimension under native conditions (10°C, 70 min, 500 V) from the right to the left side. Gels were silver stained as described earlier [11,12] or used for immunoblotting (see below).

Return Gel Electrophoresis

Return gel electrophoresis was carried out according to Schumacher et al. [12] with small modifications: 20 μg nucleic acid samples containing dye markers bromphenol blue and xylene-cyanol were applied to each slot of a 5% acrylamide slab gel in TBE with 4 M urea in addition. First, electrophoresis was performed from top to bottom under native conditions (10°C, 1 × TBE, 220 V) until xylene-cyanol nearly reached the bottom of the gel, then the buffer was exchanged to a 'low salt'-buffer (1/8 × TBE) which was heated to 60°C. Second, electrophoresis was performed in reverse direction under denaturing conditions (60°C, 1/8 × TBE, 220 V) until the xylene-cyanol reached the upper border of the gel.

Hybridization Tests

Nucleic acids were transferred from the gel to a nylon membrane (Biodyne B, 0.45 μg, Pall Biosupport, Dreieich, FRG) by electroblotting (0.5 × TBE, 2 hr, 1 A) and hybridized with the following probes:

PSTVd-probe
Digoxigenin-labeled (Boehringer Mannheim, FRG), full-length transcripts in (-)-strand orientation from cDNA of potato spindle tuber viroid (PSTVd) were used. The plasmid pRH714 [13] was linearized with

*Eco*RI and transcribed for two hours with the T7 RNA polymerase in TK-buffer containing 1 mM ATP, 1 mM CTP, 1 mM GTP, 0.65 mM UTP, and 0.35 mM DIG-UTP (Boehringer, Mannheim, FRG).

CCCVd-probe

Sonicated plasmid, containing CCCVd-cDNA was purchased from Bresatec (Adelaide, Australia) and used to transform *E. coli* JM109. The unit length CCCVd-cDNA (246-basepairs) was cut from the plasmid with *Bam*HI and subcloned into the *Bam*HI-site of the transcription vector pTZ18U [14] using standard methods [15]. The new plasmid pUW34 was linearized with *Hind*III and digoxigenin-labeled full-length RNA-transcripts of CCCd-cDNA in (-)-strand orientation were generated as described above.

CbVd-probe

A digoxigenin-labeled synthetic oligonucleotide
5´-CACTGGATTCCGTTGCAGCGCTGCCAGGGAACCCAG-3´
containing a (-)-strand sequence of the so-called structural conserved central region of CbVd1 [16] was used. The digoxigenin label was added to the 3´ end of 4 µg oligonucleotide using 60 u of the terminal deoxynucleotidyl transferase (Gibco BRL, Bethesda, Maryland, USA) for six hours in buffer supplied by the manufacturer containing 0.35 mM dATP and 0.04 mM DIG-dUTP (Boehringer, Mannheim, FRG)

Hybridizations were carried out under the following conditions: Membrane-bound nucleic acids were denatured on the membrane by incubation for 15 min in 50 mM NaOH. The denatured nucleic acids were incubated in 20 × SSC (20 × SSC: 3 M NaCl, 0.3 M Na-citrate, pH 7.0) for 5 min, washed briefly with water, and attached to the membrane by ultraviolet-cross-linking for 2.5 min with about 10 mW/m^2 at 254 nm. Prehybridizations were carried out for two hours at 60°C with the PSTVd-probe, at 60°C with CCCVd-probe under stringent conditions and at 42°C under non-stringent conditions, and at 55°C with the CbVd-probe, in 25 ml/100 cm^2 membrane of a solution containing 5 × SSC, 0.1% N-lauroylsarcosine, 0.02% SDS, 5% blocking reagent (Boehringer, Mannheim, FRG), and 50% formamide. Membranes were hybridized at the temperatures listed above for 16 hr with 5 ml/100 cm^2 membrane prehybridization solution containing 40 ng/ml digoxigenin-labeled RNA transcript probe or 120 ng/ml digoxigenin-labeled oligonucleotide probe in addition. Filters were washed twice at room temperature in 2 × SSC, 0.1% SDS for 5 min and twice at 60°C with the PSTVd-probe, at 65°C with the CCCVd-probe under stringent conditions and at 42°C under non-stringent conditions, and at 60°C with the CbVd-probe, in 0.1 × SSC, 0.1% SDS for 15 min. The digoxigenin-labeled probes were detected immunologically with slight modifications according to the supplier's manual (Boehringer, Mannheim, FRG, Application Manual: DNA Labeling and Detection Nonradioactive). The wet filters were washed briefly with PBS containing 0.1% Triton X-100, and free binding sites were saturated by incubating the membranes for 30 min in PBS containing 0.5% blocking reagent, 50 µg/ml denatured herring sperm DNA and 0.2% NaN_3. Digoxigenin-labeled probes were identified by incubation for 30 min with a 1:5000 diluted alkaline phosphatase conjugated anti-digoxigenin antibody, supplied by the manufacturer, in PBS containing 1% bovine serum albumin and 0.2% NaN_3. Filters were washed twice with PBS containing 0.1% Triton X-100 for 15 min each, and the hybridization signals were visualized by staining with 5-bromo-4-chloro-3-indolyl phosphate toluidinium salt (BCIP, 175 µg/ml) and 4-nitroblue tetrazolium salt (NBT, 340 µg/ml) in AP-buffer as described [17].

Reference viroids

Highly purified PSTVd and CCCVd RNA as well as nucleic acid crude extracts from coconut palms containing CCCVd (both samples were obtained from Dr. J. Randles, Adelaide, Australia) and from *Coleus blumei* containing CbVd (obtained from Dr. H.L. Sänger, Munich, FRG) were used as references.

Detection of dsRNAs on Immunoblots by Using a dsRNA-specific Monoclonal Antibody

Nucleic acids were transferred after electrophoresis to the positively charged Biodyne B membrane (0.45 µm, Pall Biosupport, Dreieich, FRG) by electroblotting (1 A, 2 hr). When separation was carried out by return gel electrophoresis, gels were incubated at 10°C for 15 min in order to allow renaturation of double-stranded structures before blotting. After transfer, free binding sites were saturated by incubating the membrane in PBS, containing 0.5% blocking reagent (Boehringer, Mannheim, FRG), 50 µg/ml denatured herring sperm DNA, and 0.2% NaN_3. Membrane-bound dsRNAs were identified by using 1:4 diluted supernatant of the J2 hybridoma-line, which was prepared in our laboratory and produces a dsRNA-specific IgG 2a antibody [18]. After incubation at room temperature for 1.5 h the filter was washed twice each for 15 min with PBS containing 0.1% Triton X-100. Bound J2 was detected by incubation for 45 min with alkaline phosphatase conjugated goat-ant-mouse IgG (H+L) (Dianova, Hamburg, FRG). After washing as above dsRNA-bands were visualized by staining with 5-bromo-4-chloro-3-indolyl phosphate disodium salt (BCIP, 175 µg/ml), and

4-nitroblue tetrazolium (NBT, 340 µg/ml) in AP-buffer as described [17]. Less than 50 pg dsRNA/band can be detected; unspecific cross-reactivity with PSTVd does not occur up to 100 ng viroid RNA/band [18]. The double-stranded character of dsRNA was confirmed according to Rogers et al. [19] by using RNase A, DNase I, and S1-nuclease.

RESULTS AND DISCUSSION

Search for Viroids by Electrophoretic Methods and by Viroid-specific Hybridization

Singh and coworkers [3] had reported that RNAs with viroid-like electro-phoretic properties were contained in nucleic acid extracts of oil palms from a location with occurrence of fatal yellowing. The study was started to clarify whether viroids might be the causative agent of yellowing of oil palms. Viroids may be detected by two different diagnostic principles. The first is based on their characteristic structure of a covalently closed single-stranded circle, which leads to unique gel electrophoretic properties. Consequently these features are independent upon a particular nucleotide sequence. The standard methods in this case are the two-dimensional gel electrophoresis and the return gel electrophoresis (cf. Methods). The second type of diagnosis uses for detection the nucleotide sequence of a viroid. Particularly useful is a section with a sequence, which is conserved among most viroid species. Any known characteristic sequence may be detected by molecular hybridization (cf. Methods).

Gel electrophoretic analysis
    In two-dimensional electrophoresis viroid RNA migrates outside of the diagonal trace of all other nucleic acids [10], and in return gel electrophoresis its position is behind the edge of the other nucleic acids [12]. When nucleic acid extracts of oil palms were analyzed by return gel electrophoresis, a group of at least eight RNA bands was observed in the region expected for viroids. These bands together with PSTVd in control slots are shown in Fig. 2.1. This experimental finding is in agreement with the report of Singh et al. [3], however, the number of 'viroid-like' RNAs was found to be even higher than reported before. That might be due to the larger amount of nucleic acids applied to the gel or due to more favorable conditions chosen for the run under denaturing conditions during return gel electrophoresis. As discussed in the original description of return gel electrophoresis [12] the method is efficient for a routine test of many samples. Circularity and size of the RNA under investigation should be determined before. It means that an unequivocal

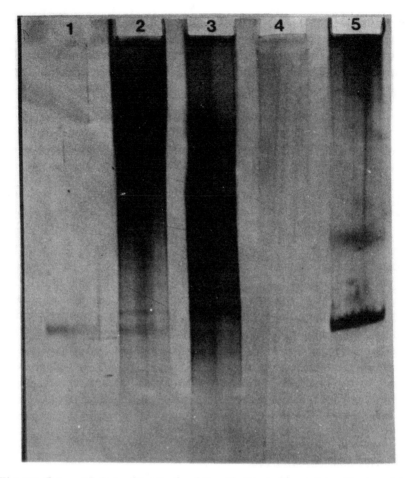

Fig. 2.1. Return gel electrophoresis of nucleic acid extracts: The nucleic acids were stained with silver. Lane 1: purified PSTVd; lane 2: a mixture (1:1) of nucleic acid extracts from PSTVd-infected tomato leaves and one oil palm leaf; lane 3: nucleic acid extract from one oil palm leaf; lane 5: nucleic acid extract from PSTVd-infected tomato leaves; The position of PSTVd in lane 1 and 5 is clearly different from those of the viroid-like bands of nucleic acid extracts from oil palm (lanes 2 and 3).

identification of a circular RNA structure cannot be achieved by return gel electrophoresis, but, only by two-dimensional gel electrophoresis.

The result of such an analysis is shown in Fig. 2.2a. Though a group of several nucleic acid bands was observed outside of the diagonal trace, neither the number of these bands nor their positions were typical for viroids. Their electrophoretic mobility under denaturing conditions in horizontal direction was markedly lower than that of PSTVd (Fig. 2.2b), and calculated on the basis of the RF-values of these bands compared

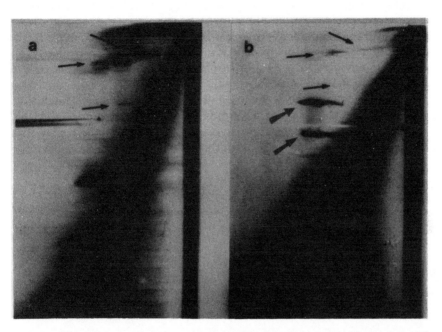

Fig. 2.2. Two-dimensional gel electrophoresis of nucleic acid extracts. The nucleic acids were stained with silver. Nucleic acid extract from oil palm leaves was analyzed on 5% PAGE (a); nucleic acid extracts from oil palm leaves was mixed 1:1 with nucleic acid extracts from PSTVd-infected tomato leaves and analyzed on 7.5% PAGE (b). The area of the oil palm RNA bands of interest is between the small black arrows and the diagonal, the PSTVd-bands (linear and circular) are designed by thick arrows. The thin arrows point to the left tip of the bands; the left and right bands were clearly visible in the original gel, but are badly visible in the photographic reproduction. The difference in intensity of the bands between the analysis in (a) and (b) is due to the variable yield of the nucleic acid extraction from different oil palm samples.

with those of the blue markers seemed to be identical with the bands observed as 'viroid-like' in the return gel (Fig. 2.1). The RNA character was established by enzymatic treatment. In summary, two-dimensional gel electrophoresis yielded a gel pattern quite untypical for viroids, leading to serious doubts about a viroid etiology of fatal yellowing.

Hybridization tests

The second type of diagnosis was performed by molecular hybridization with probes against CCCVd, PSTVd, and CbVd1. These three probes were selected for the following reasons: (1) Coconut cadang-cadang and related viroids are known to infect palm trees

[20,8,21], and are therefore of particular interest. (2) The PSTVd probe was chosen as representative for a large group of viroids containing the so-called central conserved region, which is a nucleotide sequence with a high degree of homology among all viroids of this group [22,23]. (3) CbVd1 is a viroid, found recently in *Coleus blumei* with nearly no sequence homology to the other viroids [16]. Of particular interest is the fact that CbVd1 was detected first in Brazil [24], i.e. in some geographical neighborhood of the location of fatal yellowing. The nucleic acid extracts were applied to polyacrylamide gel electrophoresis (PAGE), blotted to nylon membranes, and hybridized either with the transcripts of cloned CCCVd or PSTVd or with the CbVd1-specific oligonucleotide.

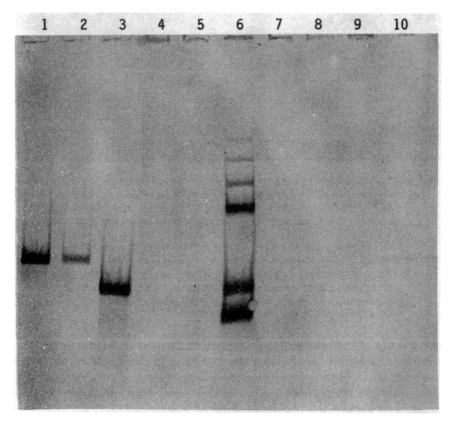

Fig. 2.3. Analysis of nucleic acid extracts from oil palms for the presence of PSTVd-like RNAs. The nucleic acids were separated on 5% PAGE, blotted, and hybridized to digoxigenin-labeled (-)-strand PSTVd-transcript. Lane 1: 10 pg PSTVd; lane 2: 1 pg PSTVd; lane 3: CCCVd; lanes 4 and 5: 50 μg nucleic acid from symptomless oil palms; lane 6: CCCVd from crude extract; lanes 7–10: 50 μg nucleic acid from diseased oil palms. No indication for a PSTVd-like sequence could be detected in the samples from oil palms.

The results together with the controls are shown in Figs. 2.3, 2.4, and 2.5. Hybridization with the PSTVd-specific probe (Fig. 2.3) exerted clear signals with PSTVd (10 pg and 1 pg), purified CCCVd, and a CCCVd-crude extract. The high sensitivity of the hybridization tests is obvious, because CCCVd could be detected clearly with PSTVd-specific probe, i.e. the central conserved region was sufficient to detect viroids containing this region. In the slots containing the samples from healthy as well as diseased oil palms, however, no indication for a signal above background could be detected. Thus, viroids with sequence homology to PSTVd could not be detected in oil palms.

The CCCVd-specific probe was applied to search for viroids with CCCVd-related sequences. In Fig. 2.4a it can be seen that none of the

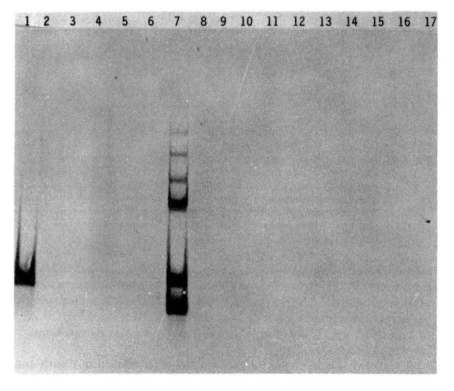

Fig. 2.4a. Analysis of nucleic acid extracts from oil palms for the presence of CCCVd-like sequences. The nucleic acids were separated on 5% PAGE, blotted, and hybridized to digoxigenin-labeled (-)-strand CCCVd-transcript. Hybridization was carried out under stringent conditions. Lanes 1 and 7: CCCVd; lanes 2–6: nucleic acid extract from symptomless oil palms (20 μg); lanes 8–17: nucleic acid extract from diseased oil palms (20 μg).

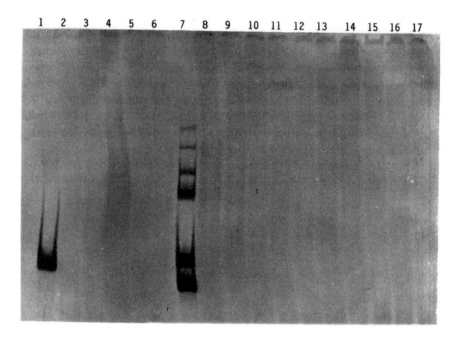

Fig. 2.4b. Analysis of nucleic acid extracts from oil palms for the presence of CCCVd-like sequences. The nucleic acids were separated on 5% PAGE, blotted, and hybridized to digoxigenin-labeled (-)-strand CCCVd-transcript. Hybridization was carried out under low stringency conditions: Lanes 1 and 7: CCCVd; lanes 2-6: nucleic acid extract form symptomless oil palms (20 µg); lanes 8–17: nucleic acid extract from diseased oil palms (20 µg). No indication for a CCCVd-like sequence could be detected in the oil palm samples.

oil palm samples contained such a sequence. In this case hybridization conditions of high stringency as well as those of low stringency were used. Even under low stringency (Fig. 2.4b) only a weak and unspecific background could be detected. Therefore, CCCVd-related viroids had to be excluded as a detectable viroid in the oil palm samples under investigation.

The third series of hybridization tests was carried out with the CbVd1-specific probe (Fig. 2.5). Also with this probe no signal could be obtained in the samples of oil palms, whereas the three different CbVd-species resulted in clear signals.

In summary, on the basis of the hybridization experiments the presence of viroids in the oil palms under study is very improbable. This statement relates to viroids with sequence homology to PSTVd, CCVd, and CbVd.

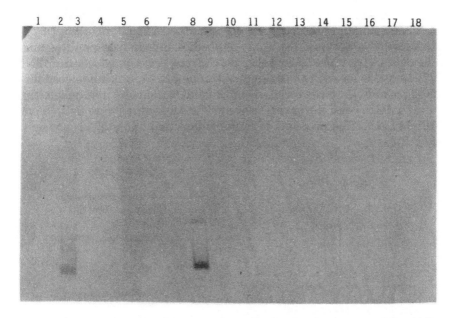

Fig. 2.5. Analysis of nucleic acid extracts from oil palms for the presence of CbVd-like sequences. The nucleic acids were separated on 5% PAGE, blotted, and hybridized to digoxigenin-labeled (-)-strand CbVd-oligonucleotide: Lane 1: nucleic acid extract from healthy *Coleus blumei* (10 μg); lane 2: nucleic acid extract from a *Coleus blumei* which contains CbVd1 and CbVd2; lanes 3–7: nucleic acid extracts from symptomless oil palms (20 μg); lane 8: nucleic acid extract from a *Coleus blumei* which contains CbVd1 and CbVd3; lanes 6–18: nucleic acid extracts from diseased oil palms (20 μg). No indication for a CbVd-like sequence could be detected in the samples from oil palms.

Detection of Double-stranded RNA

Since it had been found that the nucleic acid extracts from oil palm were free of viroids, our interest concentrated on the RNAs which appeared 'viroid-like' in return gel electrophoresis (cf. Fig. 2.1) and to a much lesser extent in two-dimensional gel electrophoresis (cf. Fig. 2.2). Their migration demonstrated that these molecules were highly retarded under denaturing conditions, as characteristic for circular nucleic acids. This behavior might also be explained, however, assuming double-stranded RNAs. Such an assumption was derived from the fact that under conditions denaturing viroids, dsRNA might be denatured only partly, i.e. not separated into its single strands. Such structures, however, migrate as slowly as denatured viroids [25] and therefore might lead to a misinterpretation of return gel electrophoresis results.

Immunoblotting experiments

The presence of dsRNA in oil palms could be tested with monoclonal antibodies against dsRNA, recently obtained in this laboratory by N. Lukács and coworkers [18]. These antibodies, prepared originally against the dsRNA from the killer yeast *Saccharomyces cerevisiae* strain YNN 27, do not react with DNA, single-stranded RNA, or viroids; less than 50 pg of dsRNA, however, is specifically detectable by immunoblotting after PAGE, even if the excess of nucleic acids other than dsRNA was 10^6-fold [18]. The principle of immunodetection of dsRNA is shown in Fig. 2.6.

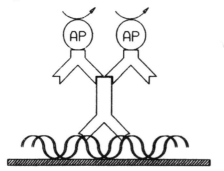

Colour reaction

Alkaline phosphatase
(Fab')$_2$ - GAM IgG

dsRNA spec. MAB I gG

dsRNA

positively charged
membrane

Fig. 2.6. Principle of immunodetection of double-stranded RNA.

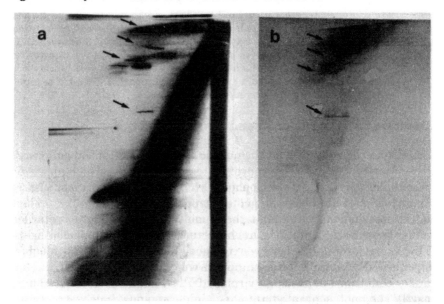

Fig. 2.7. Two-dimensional gel electrophoresis with nucleic acid extracts from oil palm: (a) silver-stained gel, (b) detection of dsRNA by immunoblotting. The arrows point to the dominant RNA species which are detected with both methods.

It is demonstrated in Fig. 2.7 that the group of oil palm nucleic acids, which were located outside the diagonal in two-dimensional gel electrophoresis, could be identified as dsRNA by immunoblotting. One should note that some dsRNA-bands could not be immunostained, most probably because they did not renature sufficiently in the second run

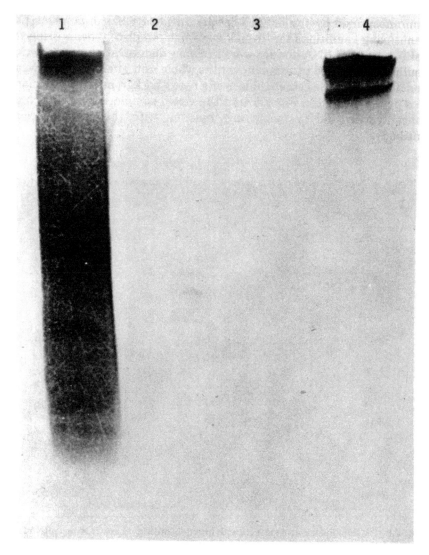

Fig. 2.8. Return gel electrophoresis with immunodetection of dsRNA: Lane 1: 50 µg nucleic acid extract from oil palm; lane 2: 50 ng PSTVd; lane 3: 50 ng CCCVd; lane 4: 10 ng reovirus dsRNA. It is shown that immunodetection is specific for dsRNA, and that the staining in the sample from oil palm is from dsRNA.

of two-dimensional gel electrophoresis. This effect was most obvious with one heavy band close to the third arrow. Only bands outside the diagonal were stained by immunoblotting, whereas the common excess of nucleic acids in the diagonal was suppressed completely. The same conclusion could be drawn from return gel electrophoresis as shown in Fig. 2.8. The bands which were called originally 'viroid-like' (cf. Fig. 2.1) were immunostained by the dsRNA-specific antibody. Since, however, the denaturing conditions for viroids were not completely denaturing for dsRNA, some dsRNA-species were in a slow denaturation-renaturation equilibrium and led to a smear over the whole slot. Therefore, only weak and diffuse bands became visible over a heavy background. Furthermore, it is demonstrated in Fig. 2.8 that the viroid sample, although in high concentration, did not yield any reaction with the dsRNA-specific antibody.

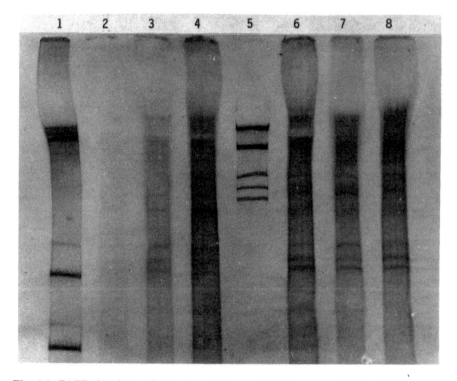

Fig. 2.9. PAGE of nucleic acid extracts with immunodetection: Lane 1: 10 ng dsRNA from *Ophiostoma ulmi;* lanes 2–3: nucleic acids extracts from symptomless oil palms from Brazil (50 µg); lane 4: nucleic acid extract from oil palms in Costa Rica free from fatal yellowing (50 µg); lane 5: 10 ng reovirus dsRNA; lanes 6–8: nucleic acid extracts from diseased oil palms from Brazil (50 µg).

Are dsRNA related to fatal yellowing?
In order to examine whether any correlation could exist between the presence of dsRNA and symptoms of fatal yellowing, all oil palm samples were tested by immunoblotting after PAGE. As the example of Fig. 2.9 shows, dsRNA was present in all samples tested, and similar patterns of dsRNA were detected in diseased as well as in symptomless oil palms. Only differences in the total concentration of dsRNA or in the concentration of individual dsRNA bands were observed. Although only low concentrations of dsRNA were found in a few symptomless trees, in others concentrations as high as in diseased trees were detected. Therefore, no correlation between the presence of dsRNA and appearance of symptoms could be observed. It should be emphasized that the sample in lane 4 was from a region in Costa Rica, which is absolutely free of fatal yellowing; particularly this sample should be clear of dsRNA signals. On the basis of the size standards for dsRNA (dsRNA from *Ophiostoma ulmi* in lane 1 and dsRNA from bovine reovirus in lane 5) the size of dsRNAs from oil palms is between 400 and 4000 base pairs.

Origin of dsRNA in oil palm
Several examples have been reported in the literature claiming that the presence of dsRNA in plants indicated a virus infection [26, 27]. Only a few examples of plant viruses with a dsRNA-genome are known [28]. But ssRNA viruses also exhibit typical dsRNA patterns, which correspond to RF- or RI-forms. The pattern of dsRNA can be used successfully as a tool for determination of virus type and strain [27]. In other cases, however, dsRNA was suggested to be of plant origin [29, 30].

Discussing our results on the dsRNA in oil palm one cannot completely exclude viral origin of the dsRNA, but it appears rather improbable for several reasons: (1) All attempts to detect virus particles in the diseased oil palms have failed until now [6]. (2) the dsRNAs were found in samples from absolutely healthy trees as well as in those from diseased trees. Therefore, at least the presence of a pathogenic virus should be excluded. (3) The lack of symptoms is characteristic for cryptic viruses, which form the most important group of plant viruses with dsRNA genome [28]. Other properties of these viruses are their low concentration in the host, lack of transmissibility by graft or by mechanical inoculation, and a bipartite or tripartite genome. In our analyses, however, we observed up to 15 dsRNA species, and even if an infection by several cryptic viruses was assumed, those virus particles which belong to the dsRNA with the highest concentration in our analysis should have been detected by electron microscopy.

38   *Plant Diseases*

On the other hand, dsRNAs which were discussed as being of plant origin [29, 30, 31] have high molecular size of 12 to 15 kb, i.e. markedly higher than observed in the oil palm extracts. The fact that the same pattern of dsRNA was detected in the Brazilian oil palms as in healthy oil palms from Costa Rica and in healthy and diseased oil palms from Ecuador (Bernard and Dollet, personal communication) points rather more to a plant origin than to cryptic viruses. Both possibilities have to be further investigated.

Differentiation between Circular RNA and dsRNA in Gel Electrophoresis

dsRNA as well as circular RNA are drastically retarded in gel

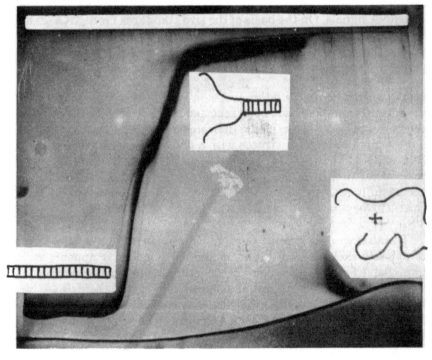

Fig. 2.10a. Temperature-gradient gel electrophoresis of the double-stranded from cucumber mosaic virus associated RNA 5 (dsCARNA5). The gel contained 5% acrylamide, 0.12% bisacrylamide, 0.08% TEMED, 8.9 mM Tris, 8.9 mM boric acid, 0.24 mM EDTA, 8 M urea, and 0.06% ammonium-peroxodisulfate for starting the polymerization. Direction of the electrophoresis was from top to bottom. Electrophoresis in the presence of the temperature-gradient was carried out at 500 V for 1.5 hr. The linear temperature-gradient was from 35°C to 65°C as indicated. The presence of the native single-stranded structure at the corresponding temperatures is indicated by schematic drawings. The experiment was described originally by Tien et al. [25].

electrophoresis due to denaturation. The main difference between the two types of RNA is that dsRNA is separated at very high temperatures into its single strands leading to an increase in electrophoretic mobility, whereas in single-stranded circular RNA a similar transition cannot occur.

The different gel electrophoretic behavior of circular RNA and dsRNA could be elucidated most easily by temperature-gradient gel electrophoresis (TGGE) [32–34]. In this technique a linear temperature gradient is established perpendicular to the direction of the electrophoresis. The sample is applied to a broad slot, which extends

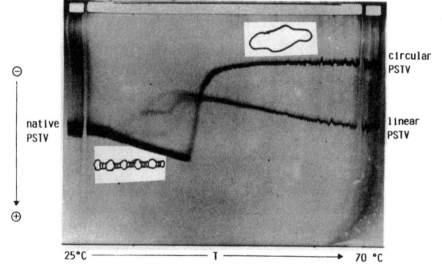

Fig. 2.10b. Temperature-gradient gel electrophoresis of PSTVd. The gel contained 5% polyacrylamide, 0.16% bisacrylamide, 0.1% TEMED, 17.8 mM Tris, 17.8 mM boric acid, 0.4 mM EDTA, and 0.07% ammonium-peroxodisulfate for starting the polymerization. Direction of the electrophoresis was from top to bottom. Electrophoresis in the presence of the temperature-gradient was carried out at 500 V for 90 min. The linear temperature-gradient is indicated. The gel was silver stained. Marker slots (left and right side of the broad sample slot) contain natural circular PSTVd and natural linear PSTVd. The experiment was described originally by Rosenbaum and Riesner [32]. The presence of the native and the denatured structure is indicated at the corresponding temperatures by schematic drawings. The linear PSTVd, which was present in the sample, contains single nicks at different sites as can be concluded from well-defined curves below the transition-temperature of the circular PSTVd.

over the whole gradient. Thus, the gel electrophoretic properties of a nucleic acid in relation to temperature can be studied within one gel electrophoretic run. In Fig. 2.10 TGGE of dsRNA (Fig. 2.10a) and PSTVd (Fig. 2.10b) are shown. Very clearly the transition during partial denaturation (two steps, at 40.5°C and 44.2°C) and the increase during strand separation (approximately 58°C) are obvious for dsRNA, whereas the mobility of circular PSTVd decreases with one dominant transition at 42°C and remains low at all temperatures above the transition. Furthermore, it could be concluded from the discontinuous transition of dsRNA at high temperature that strand separation is irreversible at low ionic strength.

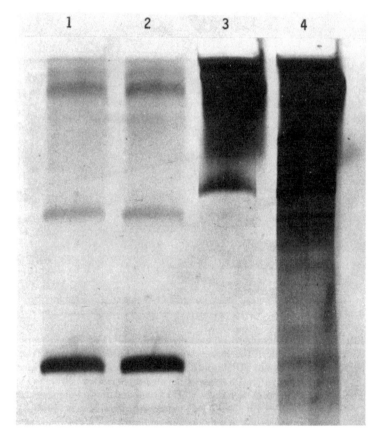

Fig. 2.11. Effect of the heating-cooling cycle on the return gel electrophoresis of viroids and dsRNA: Lane 1: 50 ng PSTVd after the boiling-cooling cycle; lane 2: 50 ng PSTVd without boiling-cooling; lane 3: 2.5 µg dsRNA from *Ophiostoma ulmi* after boiling-cooling; lane 4: 2.5 µg dsRNA from *Ophiostoma ulmi* without boiling-cooling.

Obviously, for a differentiation of dsRNA and circular RNA the different behavior at high temperatures has to be used. If the samples are boiled and subsequently cooled on ice prior to electrophoresis, dsRNA is dissociated irreversibly and remains single-stranded even at low temperature, whereas circular RNA is renatured reversibly into its native conformation. Consequently, the boiling-cooling cycle changes the behavior of dsRNA in return gel electrophoresis drastically, but does not alter the properties of circular RNA. This is shown in Fig. 2.11. Return gel electrophoresis of circular PSTVd (lane 1 and 2) leads to the same well-known pattern; return gel electrophoresis of dsRNA, however, yields the known band pattern (lane 4, similar to Fig. 2.1) only without boiling and cooling, whereas after boiling and cooling dsRNA behaves in return gel electrophoresis like a single-stranded RNA without any particular structure (lane 3). In summary, by introducing a boiling-cooling cycle one may clearly differentiate between dsRNA and circular RNA in return and two-dimensional gel electrophoresis.

ACKNOWLEDGEMENTS

We are indebted to Prof. Dr. H.L. Sänger and Mr. R.L. Spieker, Munich, for providing a sample of CbVd and making available the sequence prior to publication, and we thank Ms. B. Scholle for her skillful technical assistance, Ms. S. Fanselow and Mr. A. Kloske for their help with the experiments, and Mr. J. Oberstraß for stimulating discussions concerning the immunostaining procedure of dsRNA.

This work was supported by grants from the Deutsche Forschungsgemeinschaft, the Fonds der Chemischen Industrie and Dendé do Pará S.A. Belém—Brazil.

REFERENCES

1. Hartley, C.W.S. *The Oil Palm.* London, New York: Longman, 1988.
2. Turner, P.D. *Oil Palm Diseases and Disorders.* Kuala Lumpur: Oxford Univ. Press, 1981.
3. Singh, R.P., A.C. Avila, A.N. Dusi, A. Boucher, D.R. Trindade, W.G. van Slobbe, S.G. Ribeiro, and M.E.N. Fonseca. Association of viroid-like nucleic acids with the Fatal Yellowing Disease of oil palm. *Fitopatol. Brazil,* 13, 392–394, 1988.
4. Renard, J.L., and G. Quillec. Les maledies graves due palmier à huile en Afrique et en Amerique du Sud. *Oleagineux,* 39, 57–67, 1984.
5. Lande, H.L. van de. Diseases of fatal character to the oil palm in Suriname and in North-Brazil. *De Sur. Landb.,* 34, 15–33, 1986.
6. Slobbe, W.G. van. Spearrot (Pudrición del Cogollo). Proceedings in the IV Mesa Redonda Latino-Americana sobre Palma Aceitera; Valledupar, Colombia, June 1986, FAO, Santiago (Chile): 110–112, 1987.
7. Randles, J.W., D. Hanold and J.F. Julia. Small circular single-stranded DNA associated with foliar decay disease of coconut palm in Vanuatu. *J. Gen. Virol.,* 68, 273–280, 1987.

8. Hanold, D., and J.W. Randles. Detection of nucleotide sequences related to coconut cadang-cadang viroid in coconut and oil palm and other monocotyledons in the Pacific region. VIIIth Intern. Congr. Virol., Berlin 1990, Abstracts 93, 1990.

9. Randles, J.W. Association of two nucleic acid species with cadang-cadang disease of coconut palm. *Phytopathology*, 65, 163–167, 1975.

10. Schumacher, J., J.W. Randles and D. Riesner. A two-dimensional electrophoretic technique for the detection of circular viroids and virusoids. *Anal. Biochem.*, 135, 288–295, 1983.

11. Beuther, E., S. Köster, P. Loss, J. Schumacher and D. Riesner. Small RNAs originating from symptomless and damaged spruces (*Picea* spp.). I. Continuous observation of individual trees at three different locations in NRW. *J. Phytopathol.* 121, 289–302, 1988.

12. Schumacher, J., N. Meyer, D. Riesner and H.L. Weidemann. Diagnostic procedure for detection of viroids and viruses with circular RNAs by 'Return'-gel electrophoresis. *J. Phytopathol.* 115, 332–343, 1986.

13. Hecker, R., Z.-M. Wang, G. Steger and D. Riesner. Analysis of RNA structures by temperature-gradient gel electrophoresis: Viroid replication and processing. Gene, 72, 59–74, 1988.

14. Mead, D.A., E. Szczesna-Skorupa and B. Kemper. Single-stranded DNA 'blue' T7 promotor plasmids: a versatile tandem promotor system for cloning and protein engineering. *Protein Engineering*, 1, 67–74, 1986.

15. Maniatis, T., E.F. Fritsch, and J. Sambrook. *Molecular Cloning: A Laboratory Manual*. New York: Cold Spring Harbor Laboratory, 1982.

16. Spieker, R.L., B. Haas, Y.-Ch. Charng, K. Freimüller and H.L. Sänger. Primary and secondary structure of a new viroid 'species' (CbVd 1) present in the *Coleus blumei* cultivar 'Bienvenue'. *Nucleic Acids Res.*, 18, 3998, 1990.

17. Leary, J.J., D.J. Brigati and D.C. Ward. Rapid and sensitive colorimetric method for visualizing biotin-labeled DNA probes hybridized to DNA or RNA immobilized on nitrocellulose: Bio-blots. *Proc. Natl. Acad. Sci. USA*, 80, 4045–4049, 1983.

18. Schönborn, J., J. Oberstraß, V. E. Breyel, J. Tittgen, J. Schumacher and N. Lukács. Monoclonal antibodies to double-stranded RNA as probes in crude nucleic acid extracts. Nucleic Acids Res., 19, 2993-3000, 1991.

19. Rogers, H.J. K.W. Buck, and C.M. Brasier. Transmission of double-stranded RNA and a disease factor in *Ophiostoma ulmi*. Plant Pathol. 35, 277–287, 1986.

20. Randles, J.W., G. Boccardo and J.S. Imperial. Detection of the cadang-cadang associated RNA in African oil palm and buri palm. *Phytopathol.* 70, 185–189, 1980.

21. Boccardo, G., R.G. Beaver, J.W. Randles and J.S. Imperial. Tinangaja and Bristle Top, coconut diseases of uncertain etiology in Guam, and their relationship to cadang-cadang disease of coconut in the Philippines. *Phytopathology*, 71, 1104–1107, 1981.

22. Riesner, D., and H.J. Gross. Viroids. *Ann. Rev. Biochem.*, 54, 531–564, 1985.

23. Sänger, H.L. Viroids and viroid diseases. *Acta Horticulturae*, 234, 79–87, 1988.

24. Fonseca, M.E.N., L.S. Boiteux, R.P. Singh and E.W. Kitajima. A small viroid in *Coleus* species from Brazil. *Fitopatol. Bras.*, 14, 94–96, 1989.

25. Tien, P., G. Steger, V. Rosenbaum, J. Karper, and D. Riesner. Double-stranded cucumovirus associated RNA 5: experimental analysis of necrotic and non-necrotic variants by temperature-gradient gel electrophoresis. *Nucleic Acids Res.*, 15, 5069–5083, 1987.

26. Dodds, J.A., T.J. Morris and R.L. Jordan. Plant viral double-stranded RNA. *Ann. Rev. Phytopathol.*, 22, 151–168, 1984.

27. Valverde, R.A., J.A. Dodds and J.A. Heick. Double-stranded ribonucleic acid from plants infected with viruses having elongated particles and undivided genomes. *Phytopathology,* 76, 459–465, 1986.

28. Boccardo, G., V. Lisa, E. Luisoni and R.G. Milne. Cryptic plant viruses. *Adv. Virus Res.* 32, 171–214, 1987.

29. Wakarchuk, D.A., and R.I. Hamilton. Cellular double-stranded RNA in *Phaseolus vulgaris. Plant Mol. Biol.* 5, 55–63, 1985.

30. Valverde, R.A., S. Nameth, O. Abdallha, O. Al-Musa, P. Desjardins and A. Dodds. Indigenous double-stranded RNA from pepper (*Capsicum annuum*) *Plant Sci.* 67, 195–201, 1990.

31. Wakarchuk, D.A., and R.I. Hamilton. Partial nucleotide sequence from enigmatic dsRNAs in *Phaseolus vulgaris. Plant Mol. Biol.* 14, 637–639, 1990.

32. Rosenbaum, V., and D. Riesner. Temperature-gradient gel electrophoresis: Thermodynamic analysis of nucleic acids and proteins in purified form and in cellular extracts. *Biophys. Chem.,* 26, 235–246, 1987.

33. Riesner, D., G. Steger, R. Zimmat, R.A. Owens, M. Wagenhöfer, W. Hillen, S. Vollbach and K. Henco. Temperature-gradient gel electrophoresis of nucleic acids: Analysis of conformational transitions, sequence variations, and protein-nucleic acid interactions. *Electrophoresis,* 10, 377–389, 1989.

34. Riesner, D., K. Henco and G. Steger. Temperature-gradient gel electrophoresis: A method for the analysis of conformational transitions and mutations in nucleic acids and proteins. *Adv. Electrophoresis,* Vol. 4, 169-250, 1991.

Chapter 3

PLANT VIROIDS IN POLAND

Selim Kryczyński

Selim Kryczyński is a Professor of Plant Pathology at Warsaw Agricultural University, Poland. He received his MSc in 1963 and PhD in 1969 in Plant Pathology at Warsaw Agricultural University, and has been working there since 1964. Research interests: transport of viruses and viroids in infected plants, methods of detecting infection. Address: Department of Plant Pathology, Warsaw Agricultural University, Nowoursynowska 166, 02-766 Warsaw, Poland.

CONTENTS

INTRODUCTION

One disease of plants caused by a viroid had been observed in Poland long before the term 'viroid' was introduced by Diener [1] to plant pathology. Potato spindle tuber disease symptoms have been noted for several years, and viroid etiology of the disease was confirmed in the late 1970s, although it was not officially recognized until 1986 [2]. Potato spindle tuber viroid (PSTV) is the only plant viroid which has been found so far in Poland.

Two Polish post-doctorates working at the Department of Plant Pathology of Cornell University, Ithaca, New York, took part in the first steps of research on chrysanthemum chlorotic mottle viroid (ChCMV) in the United States. The optimum temperature conditions for ChCMV symptom expression were found at 24°C, and the unfavorable influence of low temperature on symptom expression was in good accordance with the known preferences of other plant viroids [3]. The sap-transmission of ChCMV was achieved by using precooled glassware and buffers, and by adding 1% bentonite to the extraction buffer. None of 15 plant species belonging to six botanical families became infected with ChCMV [4, 5]. The stabilizing effect of high pH, bentonite, and phenol treatments indicated that a free nucleic acid infectious agent was involved. RNase but not DNase decreased the infectivity, and this infectivity was not protected by adding Mg^{++} to the inoculum, which indicated that the infectious agent was ssRNA without [2] a protein coat [6].

After Kryczyński [7] had published a review paper on plant viroids in 1976, two cooperating research groups—one at the Department of Plant Pathology of the Warsaw Agricultural University and the other at the Department of Genetics and Potato Line Breeding of the Institute for Potato Research at Młochiw—started the research program on this new group of plant pathogens, with special attention to PSTV. Later, a research group at the Institute of Biochemistry and Biophysics of the Polish Academy of Science joined this program, and the research on PSTV was started also at the Department of Virus Diseases and Potato Seed Production of the Institute for Potato Research at Bonin.

INOCULATION OF PLANTS WITH VIROIDS AND VIROID DETECTION IN PLANTS BY BIOASSAY

We attempted first to establish the most efficient method of inoculating plants with viroids. Tissue implantation with surgical cannulas proved very convenient for inoculating various plant species with different viroids [8, 9]. We assumed that plant viroids are small enough to pass through the narrow plasmodesmata of callus cells surrounding the tissue implanted into the stems of inoculated plants. It is still possible, however, that viroid infection delays callus formation, since it is commonly known that chrysanthemum stunt viroid (CSV) and ChCMV delay rooting of chrysanthemum cuttings [10]. Furthermore, Pietrak [11] has found that necrosis forms at the union-graft places of plants infected with PSTV and a high percentage of scions are not accepted. Tissue implantation proved to be as efficient as grafting in transmitting PSTV, CSV and ChCMV [5, 12] (Table 3.1) but this inoculation technique is rather laborious. It can be used exclusively for inoculating plants old enough to accept the implants without being damaged.

Table 3.1. The efficiency of transmission of three plant viroids by tissue implantation with surgical cannulas

Viroid	Source plant	Inoculated plant	Number of plants	
			Inoculated	infected
Chrysanthemum stunt, CSV	Chrysanthemum Blanche	Chrysanthemum Blanche	200	200
Chrysanthemum chloro-tic mottle, ChCMV	Chrysanthemum Deep Ridge	Chrysanthemum Deep Ridge	200	200
Potato spindle tuber, PSTV	Tomato Rutgers	Tomato Rutgers	240	227

Mechanical inoculation has been almost exclusively used for experimental and routine work with plant viroids in Poland. The inoculum was prepared by grinding leaf tissue of the source plants in a precooled mortar with 1% bentonite suspension in 0.1 M K_2HPO_4, pH = 9.3 [5, 12]. In some experiments the inoculum was additionally clarified and concentrated according to Morris and Smith [13] but we do not feel that this is necessary. 'Dry inoculation' (direct leaf-to-leaf inoculation by rubbing) was not superior to sap inoculation although in mass inoculation of potato plants with PSTV good results were obtained when inoculated plants were switched with the haulm of diseased plants. Kaczmarek [14] reported a high infection level of potato plants with PSTV when tuber germs were punctured with a needle and then soaked with inoculum. Sap-inoculation and mechanical tuber inoculation gave inferior results.

The reaction of *Scopolia sinensis* plants to infection with PSTV was not consistent [5, 12]. The local symptoms were seen only in some of the inoculated plants while the number of plants reacting with systemic symptoms was a little higher. Nevertheless, some of the inoculated *S. sinensis* plants did not react to PSTV inoculation, irrespective of whether severe (s-PSTV) or mild (m-PSTV) isolates were used.

Rutgers and Sheyenne tomato plants proved the best indicator plants for PSTV detection [5, 12], and Rutgers was used in most of the experimental and routine work. These plants reacted to PSTV infection with stunting, leaf epinasty (Fig. 3.1) and necrosis of stems (Fig. 3.2), leaf-petioles, and veins. Necrosis was more typical for s-PSTV, while 'albinism' (Fig. 3.3) was more pronounced in plants inoculated with m-PSTV. Inoculated tomato plants reacted with albinism, especially when very high temperature was prevalent during the incubation period,

Fig. 3.1. Epinasty of PSTV-infected Rutgers tomato leaf and viroid-induced necroses on leaf petiole and leaf veins.

Fig. 3.2. Necroses on the stem and leaf petiole of PSTV-infected Rutgers tomato plant.

and when inoculated plants were incubated at high light intensity and long day conditions. It is worth mentioning that in such conditions albinism is often observed also in PSTV-free tomato plants. The best time for inoculating Rutgers tomato plants with PSTV is when the plants are in a second leaf stage. The results of inoculating plantlets in cotyledon stage were erratic, and when older plants were inoculated

Fig. 3.3. Albinism of tomato cultivar Rutgers caused by PSTV infection.

many of them did not react with symptoms [5, 12]. The reaction of older plants could be improved by trimming them at 10–14 days following inoculation. In such conditions symptoms could be visible in lateral shoots.

We were not able to confirm the results indicating that manganese nutrition influences PSTV symptom development in Rutgers tomato plants [15] no matter how many manganese treatments were applied at various times following inoculation [5, 12]. On the other hand, we have fully confirmed that high temperature (25–30°C) and high light intensity (6,000–10,000 lux) are essential for good symptom expression. The critical time for the influence of temperature and light on PSTV symptom expression in Rutgers tomato plants was between the 10th and 20th day following inoculation. In some experiments the critical time was the second week following inoculation.

In optimum conditions (inoculation of Rutgers tomato plantlets in second leaf stage, incubation at high temperature and high light intensity) PSTV symptoms were usually obtained at 14–20 days following inoculation. For m-PSTV the double-inoculation technique [16] was successfully used and second inoculation was applied at the 14th to 17th day following the first one [5, 12].

Rutgers tomato plants proved to be good indicator plants for detecting CSV and cucumber pale fruit viroid (CPFV) infection [17, 18]. Potato spindle tuber viroid, CSV, and CPFV produced also very obvious symptoms in Bonnie Jean chrysanthemum plants [17, 18]. On the other hand, ChCMV infected only chrysanthemum plants and produced diagnostic symptoms exclusively in Ridge chrysanthemums [5, 17, 18].

OTHER METHODS OF VIROID DETECTION IN PLANTS AND THE PRACTICE OF PLANT VIROID DETECTION WITH VARIOUS METHODS

Polyacrylamide gel electrophoresis (PAGE) technique was also routinely used. In early experiments we used the RNA extraction method proposed by Morris and Smith [13]. Slab-gel technique proposed by Schumann et al. [19] was more convenient than tube-gel electrophoresis (Fig. 3.4). Later, we adopted the RNA extraction method described by Čech [20] using polyethylene glycol (PEG) for RNA precipitation. This method was improved by Skrzeczkowski et al. [21], who used two-step PEG precipitation. In the first step, 12% PEG 6000 is used to precipitate high molecular RNAs, and then 20% PEG 6000 is added to the supernatant to precipitate the viroid-RNA. When such extracts are applied to the gel, viroid-RNA bands are most distinct and lower concentration of viroids can be detected. This technique was routinely used also for CSV and CPFV detection [17, 18].

A cDNA probe, supplied by Dr. de Vos from the Netherlands, was successfully used for PSTV routine detection. Skrzeczkowski et al. [22] synthesized their own cDNA probe complementary to specific and conservative regions of PSTV-RNA low in G. The probes were inserted into pUC-8 plasmid for cloning, and two selected clones were labeled by nick-translation with α-[^{32}P] dCTP. Unfortunately, PSTV-RNA

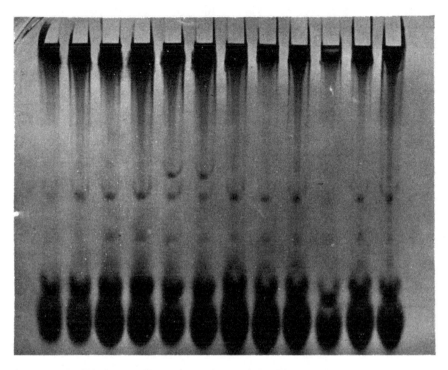

Fig. 3.4. Viroid-RNA-specific bands on 5% polyacrylamide slab gel stained with toluidine blue after electrophoresis.

hybridization with those plasmids was rather inefficient, since they were not long enough. When good cDNA probes were used, the results of dot-blot hybridization test proved to be comparable to bioassay [23], and the dot-blot hybridization test was judged sensitive and reliable enough to be used for large-scale potato testing for PSTV. The results of Kryczyński et al. [24] showed, on the other hand, that bioassay was more efficient than cDNA and especially more efficient than PAGE in detecting PSTV, since the viroid was detected in higher 'dilutions' of infected plant material with the healthy material (Table 3.2). The same cDNA probe, complementary to s-PSTV-RNA, hybridized too with CSV-RNA and CPFV-RNA, and was successfully used for detection of these two viroids in small samples of tomato seeds [25, 26].

Using bioassay on Rutgers tomatoes as indicator plants m-PSTV was detected in five to six weeks following inoculation in composed leaf samples of seven potato cultivars. On the other hand, s-PSTV was detected in some cultivars one to two weeks later or was not detected at all till the end of the testing period (Table 3.3). In secondarily infected plants m-PSTV was detected in 85% of plants while s-PSTV was detected

Table 3.2. Detection of severe (s-PSTV) and mild (m-PSTV) isolates of potato spindle tuber viroid by bioassay, PAGE or cDNA methods in a sequence of dilutions of diseased leaf tissue with healthy tissue

Tissue dilution	s-PSTV			m-PSTV		
diseased:healthy	Bioassay	PAGE	cDNA	Bioassay	PAGE	cDNA
healthy control	–	–	–	–	–	–
1:1000	+	–	–	+	–	–
1:750	+	–	+	+	–	–
1:500	+	–	–	+	–	–
1:250	+	–	–	+	–	–
1:100	+	–	–	+	–	+
1:50	+	–	+	+	+	+
1:10	+	+	+	+	+	+
diseased undiluted	+	+	+	+	+	+

Table 3.3. The detection of mild (m-PSTV) and severe (s-PSTV) isolates of PSTV in plants of seven potato cultivars in successive weeks following primary infection

Cultivar	PSTV isolate	Detection in successive weeks following inoculation					
		4	5	6	7	8	9
Azalia	m	–	+	+	+	+	+
	s	–	+	+	+	+	+
Dryf	m	–	+	+	+	+	+
	s	–	–	–	–	–	–
Pola	m	–	+	+	+	+	+
	s	–	–	–	–	–	–
San	m	–	+	+	+	+	+
	s	–	–	+	+	+	+
Sokół	m	–	–	+	+	+	+
	s	–	–	–	–	–	–
Tarpan	m	–	+	+	+	+	+
	s	–	+	+	+	+	+
Uran	m	–	+	+	+	+	+
	s	–	–	–	+	+	+

only in 58% of plants (Table 3.4), although some of the plants in which the viroid had not been detected were showing obvious symptoms. Both PSTV isolates were easily detected in various plant parts–tubers, eyes, sprouts, and leaves from different nodes and different stems [27] (Table 3.5). The lower detectability of s-PSTV compared to m-PSTV was

Table 3.4. Detection of mild (m-PSTV) and severe (s-PSTV) isolates of PSTV in plants of seven potato cultivars in secondary infection stage

Cultivar	Number of plants in which PSTV was detected versus number of plants tested		
	m-PSTV	s-PSTV	Total
Azalia	5/8	5/8	10/16
Dryf	5/6	6/8	11/14
Pola	6/8	0/8	6/16
San	8/8	6/8	14/16
Sokół	7/7	4/7	11/14
Tarpan	6/8	6/8	12/16
Uran	8/8	5/8	13/16
Total	45/53	32/55	77/100

Table 3.5. Detection of m-PSTV and s-PSTV in different plant parts of seven potato cultivars in secondary infection stage

Cultivar		Leaves	Tuber flesh	Eyes	Sprouts
m-PSTV	Azalia	5/8*	+	1/5*	nt
	Dryf	5/6	+	4/4	nt
	Pola	6/8	+	6/6	nt
	San	8/8	+	6/7	12/12
	Sokół	7/7	+	9/9	27/30
	Tarpan	6/8	+	6/6	10/10
	Uran	8/8	+	5/5	15/15
s-PSTV	Azalia	5/8	+	6/6	nt
	Dryf	6/8	−	6/6	nt
	Pola	0/8	−	0/6	nt
	San	6/8	+	14/14	3/8
	Sokół	4/7	+	1/8	20/20
	Tarpan	8/8	+	8/8	25/25
	Uran	5/8	+	4/4	2/6

*Number of plants in which PSTV was detected versus number of plants tested
nt: not tested

also noted by Kowalska-Noordam et al. [28]. Kowalska et al. [29] noted that the viroid was not easily detected in plants showing mild or no symptoms. It is very likely that lower detectability is caused by the lower replication rate of s-PSTV-RNA and lower concentration of a severe isolate in potato plants [27]. This assumption is supported by the fact that PSTV is better detected in potato plants which had been grown at high temperature conditions [30] and by the observation that in routine testing with PAGE technique m-PSTV-RNA band is usually more distinct than s-PSTV-RNA band.

REACTION OF PLANTS TO VIROID INFECTION

A thorough study of various potato genotypes reactions to infection with PSTV has been conducted at the Department of Genetics and Potato Line Breeding of the Institute for Potato Research at Mlochow [28–34] (Table 3.6). Usually no shoot or leaf symptoms were observed in the year of infection, except the growth inhibition. On the other hand, some genotypes reacted with tuber symptoms in the year of infection (primary infection). Typical shoot and leaf symptoms of secondary infection were as follows: stems showing upright growth and stunting, leaves growing at an acute angle, and leaf petioles curving to a sickle form; the neighboring leaflets were sometimes knitted (fused) they were often dark-green and rugose, and their edges were purpled and necrotic (Figs. 3.5, 3.6). The tubers from diseased plants were spindle-, pear-, or doll-shaped, small and less numerous, showing conspicuous cracking and the prominence of the eyebrows (Fig. 3.7).

The intensity of symptoms was correlated with the viroid isolate, potato genotype, and temperature. The variation of symptoms from symptomless or mild to severe was observed in genetically uniform plants grown in the same environment conditions and inoculated with s-PSTV [29], and in 29% of plants inoculated with s-PSTV the mild viroid isolate was detected [28]. About 10% of tested cultivars and potato clones showed some resistance to PSTV infection. Less than 50% of inoculated plants of cultivars Pola, Sokół, Sowa, Elida, and Jaśmin were infected, and the plants of cultivars Pola and Sokół usually reacted with mild symptoms showing some tolerance to PSTV infection [28, 34]. Some of these results were confirmed in a limited experiment [27], especially the difficulties in PSTV detection in cultivars Pola, Sokół, and Dryf. Total reduction of tuber yield varied from 25 to 63% depending on viroid isolate and potato genotype [28].

The typical reaction of chrysanthemum plants cultivar Mistletoe to CSV, and cultivar Deep Ridge to ChCMV, as well as cucumber plants cultivar Sporu to CPFV was recorded by Kryczyński and Paduch-Cichal

Table 3.6. Reaction of 10 potato cultivars to mild (m) and severe (s) strains of PSTV (second year after inoculation; potatoes grown in the field)

		Potato cultivars									
		Pola	Sokót	Atol	Azalia	Ronda	Irys	Uran	San	Tarpan	Dryf
Percentage of plants in which PSTV was detected	m	56	50	83	100	100	100	100	100	100	100
	s	37	36	100	62	100	100	100	100	100	100
Percentage of plants showing foliage symptoms[1]	m	22	10	70	9	64	5	71	56	6	76
	s	20	37	50	86	82	100	100	100	100	100
Severity of symptoms on shoots[2]	m	+	+	+	+	+	+	+	+	+	+
	s	+	+	+	++	++	+++	++	+++	++	+++
Percentage of plants showing symptoms on tubers[1]	m	22	10	40	0	64	10	15	0	20	6
	s	50	56	100	57	91	82	85	25	100	100
Percentage of tubers showing symptoms[1]	m	30	25	16	0	43	1	10	0	22	2
	s	42	46	42	12	46	37	60	10	52	41
Severity of symptoms on tubers[2]	m	+	+	+	–	++	+	+	–	+	+
	s	+	++	+++	+	++	++	+++	+	++	++
Decrease of tuber yield[1] percentage of control[3]	m	ns	24	18	38	46	37	ns	27	ns	38
	s	30	25	55	82	85	56	54	62	78	91
Decrease of tuber number[1] percentage of control[3]	m	–25	23	ns	19	40	ns	–40	ns	ns	ns
	s	ns	ns	30	50	70	ns	ns	ns	43	65

[1]Only those plants in which PSTV had been detected were taken into account.
[2]–no symptoms; + mild symptoms; ++moderate symptoms; +++ severe symptoms.
[3]Values reported represent significant differences (P = 0.05) compared to healthy controls according to Student's t test; –: indicates an increase in number of tubers per plant compared to healthy control; ns: non-significant.

[17, 18]. All three viroids— PSTV, CSV, and CPFV—caused very similar symptoms in tomato plants cultivars Rutgers and Najwczesniejszy. The stunting of plants and leaf distortion was noted, and PSTV and CPFV caused severe necrosis of leaf petioles and veins. The fruit symptoms

Fig. 3.5. PSTV-infected potato plant showing leaf epinasty and distortion.

Fig. 3.6. PSTV-infected potato plant showing growth symptoms (stiff leaves, growing at acute angle to the stem and leafroll of the upper leaves).

Fig. 3.7. Tubers of potato breeding line 22/70 showing symptoms of PSTV infection (K-3—control, healthy tubers).

were very similar to those caused by tomato mosaic virus. The fruits were small and showed typical mosaic symptoms. The darkening of seeds was a specific viroid symptom [25, 26].

Cytopathological changes in the tubers of plants infected with PSTV and potato leafroll virus were recorded [35]. They were as follows: changes in the plasmatic membrane system of the phloem parenchyma cells and companion cells and formation of vesicular structures containing fine fibrous material within the basal cytoplasm, which indicates that intensive protein synthesis takes place also in the free ribosomes in the cytoplasm. These vesicular structures were enlarging and merging, forming big vacuole-like areas. The nuclear envelope was destroyed by the outer membrane protruding into large cisternae and joining with the described vesicular system.

OTHER ASPECTS OF VIROID DISEASES AND RELATIONSHIPS BETWEEN VIROIDS

The research on PSTV epidemics was started at the Institute for Potato Research at Bonin, but so far only the literature review on possible insect transmission has been published [36]. PSTV, CSV, and CPFV were found to be pollen- and seed-transmitted in tomato. It was found too that pollination with infected pollen caused the infection of the pollinated plants [25, 26] and in recent experiments of Kryczyński and his group (unpublished) it was shown that pollen inoculation of potato plants occurs early enough to allow PSTV translocation into tubers of inoculated plants.

The results of Kryczyński and Kaminska [37] showed that ChCMV transport from implanted tissue into the stems of inoculated chrysanthemum plants started at the fifth day following inoculation and that the viroid was translocated faster up than down the inoculated stem (Fig. 3.8). The transport of PSTV in implanted Rutgers tomato stem started a little later (between the sixth and the ninth day following inoculation) [8, 9]. There is a lot of information published on the final distribution of viroids in plants but the critical research on plant viroid replication and transport in infected plants has not been initiated so far.

A comparative study of four viroids has led to the conclusion that PSTV, CSV, and CPFV may be regarded as the strains of the same viroid 'species' [17, 18]. The host ranges and reactions of particular hosts to infection with all of them were quite similar. Two-directional cross-protection was noted for each viroid pair in chrysanthemum cultivar Bonnie Jean plants. The relative mobility of RNAs of these three viroids on 5% polyacrylamide gels was identical. On the other hand, ChCMV appeared to be quite a different pathogen. This viroid infected and caused symptoms only in chrysanthemum plants and did not protect Bonnie Jean chrysanthemum plants against any other viroid. ChCMV-RNA band could never be visualized on polyacrylamide gel.

Finally, viroid-free potato and chrysanthemum plants were obtained from meristem-tips cut from PSTV-infected potato plants and from chrysanthemum plants infected with CSV, ChCMV, and CPFV after six months' therapy in a growth chamber at 5°C and 16 hours daily light of 5,000 lux intensity. Chrysanthemum plants survived quite well the conditions of therapy, while potato plants grown from stem cuttings survived these conditions much worse, and potato plants grown from tubers did not survive. Plants free of PSTV were obtained from meristem tips cut from sprouts grown from potato tubers infected with s-PSTV or m-PSTV after six months' therapy at 6–7°C in the dark. The efficiency of this therapy in viroid elimination varied for different viroids and

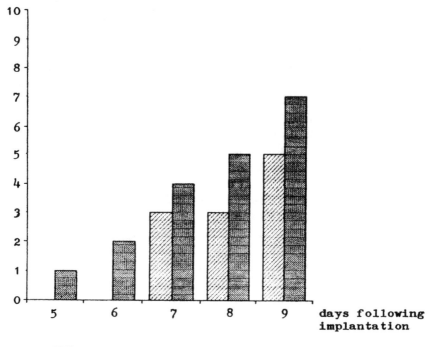

plants from cuttings taken 4 cm above
the implantation place at a given date
following inoculation

plants trimmed 4 cm below the implantation
place at a given date following inoculation

Fig. 3.8. The number of Deep Ridge chrysanthemum plants with ChCMV symptoms (versus 10 inoculated plants) from cuttings taken 4 cm above the implantation place or from plants trimmed 4 cm below the implantation place in various dates following inoculation by tissue implantation.

different plant materials from 18.5 to 80%. The low temperature therapy period could not be shortened without losing the therapeutic effect [38, 39]. Recently, promising results were obtained using chemical inhibitors (amantadine, ribavirine, and thiouracil) to eliminate viroids from infected plants.

CONCLUSION

So far, PSTV is the only viroid which has been detected in Poland. It is to be hoped that climatic conditions do not favor this pathogen; but potato growers have to be prepared to face the problem since Poland produces about 50 million tons of potatoes yearly and exports potato tuber-seeds to several European countries which have PSTV on their quarantine lists. Tomato is one of five commercially leading vegetable crops in Poland and there is danger of a viroid infection of this crop, especially under glass and plastic covers where the temperature conditions may favor the development of a viroid disease. For the same reasons, chrysanthemum and cucumber crops in greenhouses have to be protected. Hop stunt viroid and its variants recently found in fruit crops in Japan might also become a potential threat in Poland.

Polish viroid experts are prepared to diagnose plant viroid diseases. Plans are made to improve diagnostic methods by introducing return-PAGE technique and by developing and standardizing the bioassay and cDNA dot-blot test. There is a great need for studying epidemiological aspects of plant viroid diseases, possibly in connection with the identity of particular viroid pathogens and conflicting data on the role of possible vectors of plant viroids. Research is also aimed at improving the procedure of viroid elimination from diseased plants. The breeders have to be especially careful to exclude all viroid-infected materials from the plant breeding programs.

Plant pathologists studying viroid diseases are interested in the identification of particular viroids. So far, the main criterion for identification is the primary structure of viroid RNA. There appears to be a clear difference between the percentage of the structure homology of strains of a given viroid and that of unrelated viroids. Nevertheless, different viroids may cause similar diseases.

REFERENCES

1. Diener, T.O. Potato spindle tuber 'virus'. IV. A replicating low molecular weight RNA. *Virology*, 45, 411, 1971.
2. Gabriel, W., and U. Kaczmarek. Zagrożenie ziemniaków wiroidem wrzecionowatości bulw. *Ochrona Roślin*, 5/6, 3, 1986.
3. Horst, R.K., and S.P. Kryczyński. The effect of temperature on symptom expression of chrysanthemum inoculated with chlorotic mottle virus. *Phytopathology*, 61, 895, 1971.
4. Kryczyński, S.P., R.K. Horst and A.W. Dimock. Some properties of chrysanthemum chlorotic mottle virus *Phytopathology*, 61, 899, 1971.
5. Kryczyński, S., and A. Stawiszyńska. Some methods of detecting viroids in plants by infectivity assay. *Tag.-Ber. Akad. Landwirt.-Wiss. DDR Berlin*, 184, 431, 1980.
6. Horst, R.K., C.M. Geissinger and M. Staszewicz. Treatments that improve mechanical transmission of chrysanthemum chlorotic mottle virus. *Acta Horticulturae*, 36, 59, 1974.
7. Kryczyński, S. Wiroidy jako nowa grupa czynników chorobotwórczych roślin. *Postepy Nauk Rolniczych*, 3/76, 67, 1976.
8. Kryczyński, S. Przemieszczanie wiroidów i wirusów w tkankach roślin w świetle badań nad ich przenoszeniem metodą implantacji tkanki. Zeszyty Naukowe SGGW-AR, *Rozprawy Naukowe*, 119, 71 pp. + 28 figs., 1979.

64   Plant Diseases

9. Kryczyński, S. Transmission of viroids and viruses by tissue implantation and transport across the callus barrier. *Phytopath. Z.*, 106, 63, 1983.

10. Horst, R.K., R.W. Langhans, and S.H. Smith. Effects of chrysanthemum stunt, chlorotic mottle, aspermy and mosaic on flowering and rooting of chrysanthemums. *Phytopathology*, 67, 9, 1977.

11. Pietrak, J. Wpływ obecności wiroida wrzecionowatości bulw ziemniaka na wynik szczepienia roślin. *Biul. Instytutu Ziemniaka*, 29, 33, 1983.

12. Kryczyński, S., A. Stawiszyńska, A. Kowalska, S. Skrzeczkowska, K. Szkutnicka and D. Bielecka-Pluta. Methods of detecting infection of plants with severe and mild isolates of potato spindle tuber viroid. *Ziemniak—The Potato*, 1980, 33, 1980.

13. Morris, T.J., and E.M. Smith. Potato spindle tuber disease: Procedures for the detection of viroid RNA and certification of disease-free potato tubers. *Phytopathology*, 67, 145, 1977.

14. Kaczmarek, U. Influence of various methods of inoculation on the infection of potato plants with spindle tuber viroid (PSTV) and on its detectability. *Proc. Int. Seminar Viroids of Plants and Their Detection, Warsaw*. Warsaw: WAU Press, p. 63, 1986.

15. Singh, R.P., C.R. Lee and M.C. Clark. Manganese effect on the local lesion symptom of potato spindle tuber 'virus' in *Scopolia sinensis*. *Phytopathology*, 64, 1015, 1974.

16. Fernow, K.H. Tomato as a test plant for detecting mild strains of potato spindle tuber virus. *Phytopathology*, 57, 1347, 1967.

17. Kryczyński, S., and E. Paduch-Cichal. Comparative study of four plant viroids. *Proc. Int. Seminar Viroids of Plants and Their Detection, Warsaw*. Warsaw: WAU Press, p. 19, 1986.

18. Kryczyński, S., and E. Paduch-Cichal. A comparative study of four viroids of plants. *J. Phytopathol.*, 120, 121, 1987.

19. Schumann, G.L., H.D. Thurston, R.K. Horst, S.O. Kawamoto and G.I. Nemoto. Comparison of tomato bioassay and slab gel electrophoresis for detection of potato spindle tuber viroid in potato. *Phytopathology*, 68, 1256, 1978.

20. Čech, M. Viroids in Czechoslovakia. *Zeszyty Problemowe Postepów Nauk Rolniczych*, 291, 83, 1983.

21. Skrzeczkowski, L.J., E. Okely and N. Mackiewicz. Improved PAGE and cDNA diagnosis of potato spindle tuber viroid. Paper presented at AAB Meeting, Cambridge, April 10–12, 1985.

22. Skrzeczkowski, L.J., M. Welnicki, J. Hennig, W. Markiewicz, R. Kierzek and W. Zagórski. The application of synthetic DNA probe for the detection of the potato spindle tuber viroid (PSTV). *Proc. Int. Seminar Viroids of Plants and Their Detection, Warsaw*, Warsaw: WAU Press, p. 107, 1986.

23. Welnicki, M., L.J. Skrzeczkowski, S. Skrzeczkowska, M.Waś, W. Marczewski and W. Zagórski. Large scale comparison of dot-blot hybridization test and bioassay for detection of potato spindle tuber viroid. Paper presented at French-Polish Seminar Les Maladies Virale des Plantes, Warsaw, May 16–21, 1988.

24. Kryczyński, S., A. Stawiszyńska, and S. Skrzeczkowska. The possibilities of PSTV detection by different methods in big groups of plants. Paper presented at 22nd Conf. of Polish Virologists, Warsaw, October 19–20, 1989.

25. Kryczyński, S., E. Paduch-Cichal, and L.J. Skrzeczkowski. Transmission of three viroids by seed and pollen of tomato plants. *Proc. Int. Seminar Viroids of Plants and Their Detection*, Warsaw. Warsaw: WAU Press, p. 55, 1986.

26. Kryczyński, S., E. Paduch-Cichal, and L.J. Skrzeczkowski. Transmission of three viroids through seed and pollen of tomato plants. *J. Phytopathol.*, 121, 51, 1988.

27. Kryczyński, S., E. Paduch-Cichal, G. Dziurla, A. Stawiszyńska, and S. Skrzeczkowska. Detection of potato spindle tuber viroid in plants of seven cultivars

during the primary and secondary infection stage. *Zeszyty Problemowe Postępów Nauk Rolniczych*, 381, 95, 1988.

28. Kowalska-Noordam, A., M. Chrzanowska and S. Skrzeczkowska. Reaction of ten Polish potato cultivars to severe and mild strains of potato spindle tuber viroid. *Ziemniak—The Potato*, 1986/1987, 71, 1987.

29. Kowalska, A., S. Skrzeczkowska, M. Chrzanowska and D. Bielecka-Pluta. Reaction of plants of the potato clone PW 22/70 to potato spindle tuber viroid. *Ziemniak— The Potato*, 1980, 63, 1980.

30. Kowalska-Noordam, A., and S. Skrzeczkowska. Reakcja roślin ziemniaka rodu PW 22/70 na zakażenie silnym i stabym szczepem PSTV. *Biul. Instytutu Ziemniaka*, 31, 29, 1984.

31. Kowalska, A., and S. Skrzeczkowska. Reaction of potato plants infected with potato virus M (PVM) or leafroll virus (PLRV), additionally with the potato spindle tuber viroid (PSTV). Paper presented at 14th Conf. Polish Plant Virologists, Błażejewko, September 16–19, 1980.

32. Chrzanowska, M., and A. Kowalska. Reaction of Polish potato cultivars to the severe strain of potato spindle tuber viroid, in Abstr. 8th Trienn. Conf. EAPR, Munch, p. 167, 1981.

33. Chrzanowska, M., A. Kowalska-Noordam, H. Zagórska, and S. Skrzeczkowska. Reakcja polskich odmian ziemniaka na silny szczep wiroida wrzecionowatości bulw. *Biul. Instytutu Ziemniaka*, 31, 15, 1984.

34. Chrzanowska, M., A. Kowalska-Noordam, S. Skrzeczkowska and H. Zagórska. Reaction of Polish potato cultivars and clones to the sap inoculation with PSTV. *Proc. Int. Seminar Viroids of Plants and Their Detection*, Warsaw, Warsaw: WAU Press, p. 131, 1986.

35. Golinowski, W., and G. Garbaczewska. Cytological changes in phloem cells of potato plants infected with potato leafroll virus and potato spindle tuber viroid. *Zeszyty Problemowe Postępów Nauk Rolniczych*, 298, 63, 1984.

36. Werner-Solska, J. Przenoszenie wiroida wrzecionowatości bulw ziemniaka przez owady—w świetle literatury. *Biul. Instytutu Ziemniaka*, 29, 57, 1983.

37. Kryczyński, S., and E. Kamińska. Próby przenoszenia dwóch wirusów złocieni przez implantację tkanki za pomocą igieł punkcyjnych. *Roczniki Nauk Rolniczych*, Seria E, 6 (2), 211, 1976.

38. Paduch-Cichal, E., and S. Kryczyński. A low temperature therapy and meristem-tip culture for eliminating four viroids from infected plants. *Proc. Int. Seminar Viroids of Plants and Their Detection*, Warsaw. Warsaw: WAU Press, p. 137, 1986.

39. Paduch-Cichal, E., and S. Kryczyński. A low temperature therapy and meristem-tip culture for eliminating four viroids from infected plants. *J. Phytopathol.*, 118, 341, 1987.

Chapter 4

TROPICAL MOLLICUTE DISEASES
OF PLANTS

Karl Maramorosch

Karl Maramorosch is Robert L. Starkey Professor of Microbiology and
Professor of Entomology at Rutgers—The State University of New
Jersey, USA. He received his PhD degree at Columbia University in New
York in 1949. Faculty member, Rockefeller University 1949–1961.
Program Director of Insect Physiology and Virology, Boyce Thompson
Institute for Plant Research 1961–1974. Distinguished Professor,
Waksman Institute, Rutgers Univ. 1974–1985. Professor of Entomology
1985 to date. Leopoldina Academy, Fellow New York Academy of
Sciences, National Academy of Science, India, American
Phytopathological Society, Entomological Society of America, American
Association for the Advancement of Science, Electron Microscopy Society
of America, Harvey Society, Tissue Culture Association, International
Association Medicinal Forest Plants, Indian Virological Society.
Research interests: comparative virology, invertebrate cell culture, plant
pathology, parasitology. Address: Department of Entomology, Cook
Campus, Rutgers—The State University of New Jersey, New Brunswick,
NJ 08903, USA.

CONTENTS

INTRODUCTION

Mollicute diseases of plants include well-defined spiroplasma-induced diseases and less well-defined diseases caused by mycoplasma-like organisms (MLOs). The majority of the mollicute diseases have been reported from temperate regions. This does not necessarily imply that fewer occur in the tropics than elsewhere, but rather that there are fewer trained investigators and laboratories equipped for MLO disease studies in the tropics. This also becomes apparent when data presented in 1989 by the International Working Team on MLOs in Plants are analyzed [1]. Tables 4.1–4.3 list the numbers of described MLO diseases from tropical countries, compiled from these data. India surpassed all other tropical areas by far, having reported 96 diseases. Taiwan reported 63, Brazil 39, Australia 24, Solomon Islands 14, and Upper Volta 13. The latter two countries obviously had reports from some workers, perhaps assigned there by FAO, who described their findings recently. It would seem likely that Indonesia with its diverse and extensive agriculture and different islands and regions would actually have more than the 10 diseases described up to now, and Puerto Rico (2 diseases), Mexico (5), and many Caribbean countries most likely have many as yet unreported diseases of mollicutes awaiting further study. A recent book, in which more than 50 plant diseases associated with mollicutes have been described from China, listed many from the tropical island of Hainan [2].

The pioneering work on MLO diseases by L.O. Kunkel [3–7] was based on symptomatology and vector specificity. Until recently, biological criteria developed by him for 'yellows-type' diseases were the only means for classifying and distinguishing them. In 1988, R.E. Davis and his associates at the US Department of Agriculture in Beltsville, Maryland for the first time applied cloned DNA and RNA probes to investigate the genetic relatedness of MLOs [8, 9]. The use of nucleic acid probes now permits classification of MLO diseases, as well as early and rapid detection, replacing laborious and less effective techniques.

DISCOVERY OF PLANT MOLLICUTE DISEASES

Until 1967, it was believed that yellows-type plant diseases were caused by viruses. The discoveries reported in Japan in three papers in 1967, by Doi et al., Ishiie et al., and Nasu et al. [10–12], revealed the prokaryotic nature of the causative agents. All three papers appeared in the same issue of the same journal, following oral presentations given by the senior authors at the annual meeting of the Plant Pathology Society of Japan earlier that year. Since more plant pathologists than

Table 4.1. Reports of mollicute diseases from tropical areas

India	96
Taiwan	63
Brazil	39
Australia	24
Solomon Islands	14
Upper Volta	13
Sudan	12
Ecuador	11
China	10
Indonesia	10

Table 4.2. Mollicute diseases reported from other countries

Thailand	5
Malaysia	4
Morocco	3
Chile	2
Sri Lanka	2
Ivory Coast	2
Argentina	2
Togo	2
Japan	2
Single reports have come from Iran, the Philippines, Kenya, Oman, Bangladesh, Peru, Tanzania, Cameroon, Bolivia, Paraguay, Papua New Guinea, and Zanzibar	

Table 4.3. Mollicute diseases reported from the Caribbean area

Mexico	5
Cuba	4
Jamaica	3
Puerto Rico	2
Dominican Rep.	2
Colombia	2

entomologists were involved with studies of MLO diseases, the very important third paper, by the entomologist Nasu and his associates, which documented the presence of rice yellow dwarf disease MLOs not only in plants, but also in leafhopper vectors, unfortunately was ignored not only by Japanese plant pathologists, but also by most subsequent authors and reviewers [13]. Such an omission is inexcusable and unethical, constituting a non-scientific practice that must be exposed and strongly condemned.

At present, a clear distinction can be made between the cultivable spiroplasmas (Figs 4.1 and 4.2) and the fastidious MLOs. In many diseased plants and in a limited number of insect vectors, MLOs have been observed by electron microscopy. In other diseases symptomatology remained the main basis for suspecting an MLO etiology. Failure to transmit the causative agent mechanically, the ability to transmit it by grafting or dodder, and the temporary remission of symptoms by tetracycline but not by penicillin treatment have provided additional evidence for MLO etiology.

SYMPTOMATOLOGY

Witches' broom symptoms, associated with reduced leaf size, strongly suggest but do not prove MLO etiology in diseased plants. Breaking of bud dormancy, seed sterility, leaf epinasty, stunting, virescence of flowers, shortening of internodes, abnormal flower induction, big bud formation, phyllody, and proliferation of adventitious bud and shoots are usually characteristic of MLO infections. Unfortunately, some MLO diseases do not conform to the above symptoms and some of the 'typical' symptoms can be induced by other causes. Therefore, symptomatology can only provide a hint, but further confirmation by electron microscopy and antibiotic treatment may be necessary to confirm MLO etiology. The distinction between different MLOs is not possible by electron microscopy and can now be carried out by cloned nucleic acid hybridization probes. These probes can also accurately detect MLOs.

VECTOR TRANSMISSION

In nature, most plant mollicutes are transmitted from plant to plant by phloem-feeding insects. Finding the insect vector is often extremely difficult and working with such vectors, once they have been discovered, is costly and laborious. The classical experiments performed 55 years ago by Kunkel with aster yellows MLOs and *Macrosteles fascifrons* leafhoppers [3], as well as his studies performed with *Dalbulus maidis*, the corn stunt vector of *Spiroplasma kunkelii* [6], required many years

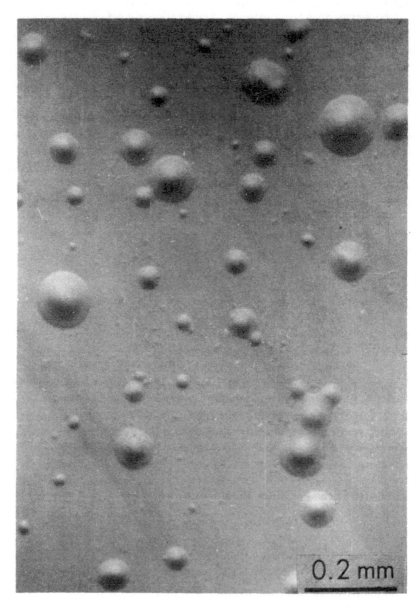

Fig. 4.1. 'Fried egg' culture of *Spiroplasma kunkelii* on agar.

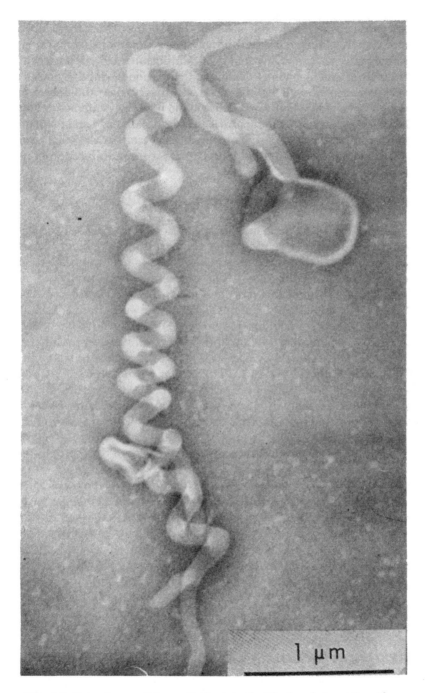

Fig. 4.2. *Spiroplasma citri*, negatively stained with phosphotungstic acid.

of very exact controlled studies, as did those of his associates L.M. Black and later K. Maramorosch and of other entomologists and plant pathologists at the University of California, US Department of Agriculture, and elsewhere. The long incubation periods in plants and insect vectors as well as the need for insectaries and insect-proof greenhouses made vector work difficult and not always attractive or even possible. Furthermore, the separation of plant pathology and entomology departments often hampered or prevented vector transmission studies of MLO agents, because plant pathologists were prohibited from using insects!

TRANSMISSION AND STORAGE

In addition to vector transmission, grafting from diseased to healthy plants of the same or related species or grafting by means of various species of dodder to related or unrelated plants has been the standard transmission procedure for MLO diseases. Interestingly, in some instances, it is possible to use dodder bridges between unrelated species, while it is not possible to transmit the MLO between plants of the same species, when using dodder as a bridge. For instance, it has not been possible to transmit aster yellows MLOs from China aster to China aster (*Callistephus chinensis*) by dodder, but it is very easy to transmit the MLOs from China aster to *Catharanthus roseus*, using *Cuscuta* sp. Grafting of monocots has presented formidable difficulties and those using dodder extensively often have found that certain species may transmit one, but not another kind of MLO. The reasons for these transmission difficulties are not well understood but they might be linked with the susceptibility of different species of dodder to different MLOs.

   Periwinkle, *Catharanthus roseus*, became the preferred storage host for mollicute diseases. Kunkel kept 18 different MLO diseases in his greenhouses in periwinkle plants for many years, but when he died in 1960, his collection unfortunately was destroyed by his successor Armin C. Braun, who was not interested in yellows-type diseases. In Canada, Chiykowski found that leafhopper vectors carrying MLOs can be frozen at $-70°C$ and used years later to inoculate stock leafhoppers so as to render them infective [14]. Other means of preserving MLOs are now being developed, as the nucleic acids extracted from diseased plants can be lyophilized and used later for serological studies.

DETECTION OF MOLLICUTES BY ELECTRON MICROSCOPY

Attempts to ascertain the presence of mollicutes by rapid negative staining and electron microscopy, as is customarily done with many plant viruses, has led to erroneous conclusions and to the identification of artifacts that were believed to be mollicutes. This error was exposed by Wolanski and Maramorosch [15] and their report put an end to the attempts to quickly identify MLOs by negative staining.

Thin section electron microscopy and the demonstration of the presence of a single, unit-type membrane and ribosomes is usually accepted as adequate proof of MLOs but the differentiation of MLOs and spiroplasmas by this technique is not always easy (Fig. 4.3) and inadequate fixation and electron microscopy have been common during the first decade that followed the Japanese breakthrough discovery of plant mollicutes.

ANTIBIOTIC TREATMENT

The most extensive tests performed with tetracycline antibiotics were reported by teams of Japanese workers shortly after the discovery of MLOs in plants. Japan's pharmaceutical industry supported work carried out jointly by several universities and experiment stations under the direction of Prof. Iida [16]. The results indicated that none of the tetracycline antibiotics could provide a permanent cure of the tested diseases, but temporary reversion of symptoms occurred when the antibiotics were taken up by the diseased plants. Studies of antibiotic uptake were made in several laboratories including our own [16b, 16c]. We found that uptake through plant roots resulted in temporary symptom remission. Drenching the soil or spraying the leaves had hardly any effect because tetracycline antibiotics were not entering the plants nor carried systemically in the treated plants. At the same time, less carefully designed experiments were published by others, who did not ascertain whether tetracyclines were taken up by treated plants and concluded erroneously that plant mollicutes have a different spectrum of antibiotic sensitivity from that of animal-infecting mycoplasmas.

Remission following tetracycline treatment but not penicillin treatment gives an indication that the causative infectious agent is not a virus but a prokaryote. Such tests, combined with electron microscopy, provide strong evidence of mollicute etiology.

USE OF CLONED DNA AND RNA PROBES
TO DISTINGUISH MLOs

Recombinant DNA methods have been developed in recent years to differentiate various MLOs. This pioneering work was carried out at the Microbiology and Plant Pathology Laboratory of the US Department

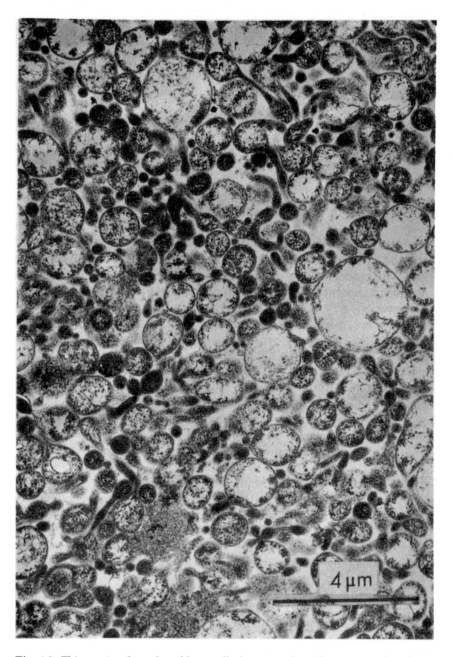

Fig. 4.3. Thin section through a phloem cell of a maize plant (*Zea mays*), infected with corn stunt spiroplasma, *S. kunkelii*. The spiroplasmas resemble typical MLOs.

of Agriculture in Beltsville, Maryland under the leadership of R.E. Davis [8]. First, a method was developed to differentiate aster yellows MLOs from other MLOs. DNA from aster yellows MLOs was enriched by removing vascular tissues of leaf midribs and incubating them with macerating enzymes at 4°C in the dark. Phloem tissue was then separated from xylem and transferred into a suspension medium containing mannitol, HEPES buffer, and polyvinyl pyrilidone at pH 7.0. Using glass homogenizers, the sieve elements were gently ruptured and the MLOs were released into the suspending medium. Centrifugation at 2000 rpm (482 xg) for 10 minutes was sufficient to remove cell organelles. The supernatant with MLOs was centrifuged for 40 minutes at 12,000 rpm (17,300 xg) and the pellets resuspended in PBS sucrose solution and stored at −70°C.

To isolate nucleic acid, the MLOs suspended in the PBS-sucrose were centrifuged at 12,000 rpm for 40 minutes and the pellet resuspended in DNA extraction buffer with the addition of/sodium dodecyl sulfate and 2-mercaptoethanol. Further incubation at 65°C, cooling, and treatment with saturated phenol and chloroform-isoamyl alcohol finally provided crude MLO nucleic acids. After further purification [9] the MLO and plant host nucleic acids were used as source of DNA for molecular cloning.

DNA and RNA riboprobes were constructed and nucleic acids were extracted from healthy plants and from different MLO diseased plants. Dot hybridization and southern blot hybridization were carried out. Of 175 recombinant plasmids, 43 were identified to react with nucleic acids from infected, but not from non-infected periwinkle plants. Several MLO-specific recombinant plasmids were used to reconstruct DNA and RNA probes and for detecting MLOs in periwinkle plants. The riboprobes can differentiate MLOs and identify them when present in different plant hosts. Thanks to this new approach, it has become feasible to establish nucleotide sequence homologies between different MLOs and study their genetic interrelatedness.

Lee et al. [17] demonstrated that non-radioactive biotin-labeled DNA preparations offer an alternative means for isolation of specific recombinants that permit the detection of specific MLO clones. Dot hybridization indicated that Italian big bud, periwinkle little leaf, and clover phyllody are closely related to tomato big bud MLO, while potato witches' broom, ash yellows, Western X, and clover proliferation MLOs are only distantly related to the tomato big bud MLO. Genetic relatedness of a European MLO and American MLOs could also be detected by dot hybridization and southern blot analysis [18]. Southern blot analyses demonstrated the occurrence of extrachromosomal DNA in MLOs of aster yellows, tomato big bud, and periwinkle little leaf [20].

LOSSES CAUSED BY MLO DISEASES

In tropical regions, losses caused by mollicute diseases can be catastrophic and several instances of large devastations are well known. We will mention here only one—caused by the lethal yellowing disease of palms. This disease has devastated coconut and other palms in some of the Caribbean islands and on the American mainland, including the southernmost part of the United States and Mexico, as well as areas in West Africa. The MLO etiology of this disease, first discovered in our laboratory [20] and later confirmed by others, could easily be established by electron microscope examination. Attempts to control the disease by tetracycline treatment have been given up and the only control measure today is the replanting with resistant palms. Recent reports of lethal yellowing susceptibility of Malayan dwarf coconut palms contradict findings made in Jamaica and a few other areas in earlier years. The degree of damage inflicted by this and other tropical mollicute diseases depends on the MLOs, the susceptibility of plant hosts to MLOs and to vectors, size of vector population, vector competence, climatic conditions, alternate plant hosts, and other factors. Some of these factors can be manipulated to provide some measure of control but others are not dependent on human intervention.

CONTROL

Chemotherapy has been attempted primarily with tetracyclines but only disease remission and no permanent cure of mollicute diseases has been achieved. Besides, chemotherapy has not been cost-efficient and possible persistence of tetracyclines in the fruit of treated trees hampered their continuous application in commercial plantings. Selected MLO diseases have been controlled by vector control [24, 25]. China asters grown commercially for seed production have been screened with cheesecloth or other netting to prevent insect transmission of aster yellows MLOs. Sugarcane white leaf disease has been treated successfully by subjecting canes to elevated temperatures and thus destroying MLOs without killing the canes. Surgery has been used for control of papaya bunchy top where the slow systemic spread of the MLOs permitted a certain degree of control. A more detailed review of non-chemical control measures has been published, [21].

Use of resistant varieties provides the most suitable control. The aim is breeding for resistance to MLOs as well as to vectors of MLOs, so as to provide plant material that can be grown in the tropics without succumbing to MLO diseases. The already mentioned Malayan dwarf coconut palms, resistant to lethal yellowing, are now being crossed with

Jamaica tall and other varieties to obtain desirable and resistant planting material. Whether resistance can be obtained and whether it will be permanent is to be seen. Among the difficulties encountered in this work is the reliance on natural infection, because no experimental transmission of lethal yellowing MLOs by means of *Myndus crudus* leafhoppers could be relied upon [22].

In south India, where the sandal spike MLO disease presents a serious problem, the only means of experimental inoculation is by grafting and no source of resistance has been found as yet. Sandal trees there are the monopoly of Karnataka State and no private ownership or private initiative in developing resistance is conceivable there. Trees growing in forests are too valuable to be sacrificed for experimental infection. Some branches could be removed at best and tested while kept alive, but such tests would have to be carried out on a large scale to be meaningful.

Breeding for insect vector resistance should be incorporated into breeding programs whenever possible [23]. Such resistance to vectors has been obtained sporadically and coincidentally, but not deliberately [26]. Other control possibilities include cross-protection as well as the production of transduced plants that carry *Bacillus thuringiensis* genes [25, 27]. There are safety aspects that have to be considered before such novel approaches can be tried in the tropics or in temperate regions.

SUMMARY

Many of the 300 known plant mollicute diseases affect tropical crops. Gene cloning and hybridoma have provided sensitive probes for assaying MLO diseases. Losses in the tropics range from barely perceivable to catastrophic, depending on the pathogens, plant susceptibility, vector population, competence, and other factors, some of which can be manipulated to provide various degrees of control. Chemotherapy has been costly and disappointing. Heat treatment, surgery, and screens to avoid vector transmission have been used in only a few instances. Breeding for resistance to mollicutes and to vectors is the current control choice. Cross-protection might be improved for selected control. New strategies will employ biotechnology to control vectors and to induce plant resistance.

ACKNOWLEDGEMENT

Much of the work reported herein that was carried out in the author's laboratory was supported for many years by USPH, NSF, FAO/UNDP, and Rockefeller Foundation Grants, as well as by the New Jersey Agricultural Experiment Station.

REFERENCES

1. McCoy, R.E., A. Caudwell, C.J. Chang, T.A. Chen, L.N. Chiykowski, M.T. Cousin, J.L. Dale, G.T.N. de Leeuw, D.A. Golina, K.J. Hackett, B.C. Kirkpatrick, R. Marwitz, H. Petzold, R.C. Sinha, M. Sugiura, R.F. Whitcomb, I.L. Yang, B.M. Zhu and E. Seemuller. *The Mycoplasmas*, Vol. V, pp. 545–640, 1989.
2. Gong, Z., Z. Chen and J. Shen. *Atlas of Plant MLO in China.* (In Chinese.) Beijing: Dwon Huang Cheng Gen, 96 pp., 1990.
3. Kunkel, L.O. Studies on aster yellows. *A.M. J. Bot.*, 13, 646–705, 1026.
4. Kunkel, L.O. Wire screen fences for the control of aster yellows. *Contrib. Boyce Thompson Inst.*, 3, 85–123, 1931.
5. Kunkel, L.O. Heat cure of aster yellows in periwinkles. *Am. J. Bot.*, 28, 761–769, 1941.
6. Kunkel, L.O. Studies on a new corn virus disease. *Arch. Gesamte Virusforschung*, 4, 24–46, 1948.
7. Kunkel, L.O. Cross protection between strains of yellows-type viruses. *Adv. Virus Res.*, 3, 251–273, 1955.
8. Davis, R.E., I.M. Lee, E.L. Dally and S.M. Douglas. Cloned nucleic acid hybridization probes in detection and classification of mycoplasma-like organisms (MLOs). *Acta Horticulturae*, 234, 115–122, 1988.
9. Lee, I.M., and R.E. Davis. Detection and investigation of genetic relatedness among aster yellows and other mycoplasma-like organisms by using cloned DNA and RNA probes. *Mol. Plant-Microbe Interactions*, 1, 303–310, 1988.
10. Doi, Y., M. Terenaka, K. Yora and H. Asuyama. Mycoplasma- or PLT group-like organisms found in the phloem elements of plants infected with mulberry dwarf, potato witches' broom, aster yellows, or paulownia witches' broom. *Ann. Phytopathol. Soc. Jpn.*, 33, 259–266, 1967.
11. Ishiie, T., Y. Doi, K. Yora and H. Asuyama. Suppressive effects of antibiotics of tetracycline group on symptom development of mulberry dwarf disease. *Ann. Phytopathol. Soc. Jpn.*, 33, 267–275, 1967.
12. Nasu, S., M. Sugiura, T. Wakimoto and T.T. Iida. On the pathogen of rice yellow dwarf virus. *Ann. Phytopathol. Soc. Jpn.*, 33, 343–344, 1967.
13. Whitcomb, R.F. and J.G. Tully (eds.) *The Mycoplasmas V. Spiroplasmas, Acholeplasmas, and Mycoplasmas of Plants and Arthropods.* Academic Press, 653 pp., 1989.
14. Chiykowski, L.N. Maintenance of mycoplasmalike organisms. In: *Tree Mycoplasmas and Mycoplasma Diseases.* C. Hiruki, ed. Univ. Alberta Press; pp. 123–134, 1988.
15. Wolanski, B., and K. Maramorosch. Negatively stained mycoplasmas: fact or artifact? *Virology*, 42, 319–327, 1970.
16a. Anonymous. Tetracycline Research Group. Effects of tetracycline antibiotics on various plants with presumptive mycoplasma diseases. Tokyo. (In Japanese.), 80 pp., March 1969.
16b. Klein, M., R. J. Frederick and K. Maramorosch. Chemotherapy of aster yellows: tetracycline hydrochloride uptake by healthy and diseased plants. *Phytopathology*, 61, 111–115, 1972.
16c. Frederick, R.J., M. Klein and K. Maramorosch 1971. Acquisition and retention of tetracycline HCl by plants. *Plant Dis. Reporter*, 55, 223–226.
17. Lee, I.M., R.E. Davis and N. DeWitt. Nonradioactive screening method for isolation of disease-specific probes to diagnose plant diseases caused by mycoplasma-like organisms. *Appl. & Environ. Microbiol.*, 56, 1471–1475, 1990.
18. Bertaccini, A., R.E. Davis and I.-M. Lee. Detection of chrysanthemum yellows mycoplasma-like organism by dot hybridization and southern blot analysis. *Plant Dis.*, 74, 40–43, 1990.

19. Davis, R.E., I.M. Lee, S.M. Douglas, E.L. Dally and N. DeWitt. Development and use of cloned nucleic acid hybridization probes for disease diagnosis and detection of sequence homologies among uncultured mycoplasma-like organisms (MLOs). *Art Zbl. Bakteriol. Hyg.* Suppl. 20, 303-307, 1990.

20. Plavsic-Banjac, B., P. Hunt and K. Maramorosch. Mycoplasma-like bodies associated with lethal yellowing disease of coconut palms. *Phytopathology*, 62, 298–299, 1972.

21. Maramorosch, K. Non-chemical control of plant mycoplasma diseases. In: *Mycoplasma Diseases of Crops: Basic and Applied Aspects.* K. Maramorosch and S.P. Raychaudhuri, eds. Springer-Verlag, pp. 431–449, 1988.

22. Tsai, J.H. and K. Maramorosch. Lethal yellowing of palms. In: *Mycoplasma Diseases of Woody Plants.* S.P. Raychaudhuri and N. Rishi, eds. New Delhi: Malhotra Publ. House, pp. 29–39, 1988.

23. Maramorosch, K. Insects and plant pathogens. In: *Breeding Plants Resistant to Insects.* F. Maxwell and P. Jennings, eds. Wiley-Interscience, pp. 137–155, 1980.

24. Maramorosch, K. Control of vector-borne mycoplasmas. In: *Pathogens, Vectors, and Plant Disease.* K.F. Harris and K. Maramorosch, eds. Academic Press, pp. 299–329, 1982.

25. Maramorosch, K. Strategies against plant mycoplasma diseases and vectors of MLOs. *Proc. 6th Auchenorrhyncha Conf.*, Turin, pp. 431–443, 1987.

26. McKelvey, J.J., Jr., B.F. Eldridge and K. Maramorosch (eds.). *Vectors of Disease Agents: Interactions with Plants, Animals, and Man.* New York: Praeger, 243 pp., 1981.

27. Maramorosch, K. (ed.) *Biotechnology for Biological Control of Pests and Vector,* CRC Press, 1991.

Chapter 5

LETHAL YELLOWING OF PALMS

James H. Tsai

James H. Tsai is a professor at the Fort Lauderdale Research and Education Center, Institute of Food and Agricultural Sciences, University of Florida. He received his BSc in Plant Pathology, National Chung Hsing University, Taiwan in 1957, MSc in 1967 and PhD in 1969 in Entomology, Michigan State University. 1973–1978 Assistant Professor, 1978–1984, Associate Professor, 1984 to date Professor, University of Florida. American Phytopathological Society, Entomological Society of America, Entomological Society of China, International Working Groups of Legume and Maize Viruses, International Organization of Citrus Virologists, International Organization of Mycoplasmology. Research interests: plant virus and prokaryote identification, isolation and characterization, pathogen-vector-host relationship. Address: Fort Lauderdale Research and Education Center, University of Florida, 3205 College Avenue, Fort Lauderdale, FL. 33314.

CONTENTS

INTRODUCTION

Among the 2,000 species of palm, the coconut, date and oil palms are important to world economy. Coconut palm (*Cocos nucifera* L.) is one of the most valuable trees in the world. Over 25 million tons of coconuts are produced each year on 6.5 million hectares of land throughout the tropics. For millions of people, this tree of life is not only a cash crop but a source of food, shelter, fiber, drink, and clothing. Many tropical islands in the Pacific are inhabitable only because of the availability of coconut palms [1], and coconut palms are also an important component of the tropical forest ecosystem of Africa and South America [2].

Most parts of the coconut tree have economic value. The wax-like inflorescences and the spathes of inflorescence are used for decorations. Coconut wood is used for tools, furniture, and building materials. Leaves are woven into screens, hats, baskets, and fans; they are also used to thatch houses. The midribs of leaves are used for brooms, and the sheath of the leaf base is woven into sandals and strainers. The shells of the nuts are used for containers and ornaments as well as for recovery of valuable activated charcoal and shell chemicals. Coconut husks are made into mats, brushes, upholstery stuffings, cordage, carpets, rope and various types of rubberized coir products.

Different parts of the tree also provide important foodstuffs. The palm hearts are eaten raw. Sap from inflorescences is used as a soft drink or can be made into wine. Coconut meat contains 20% protein and is nutritious food for humans and livestock, and it is also used for oil, ointment, and toiletries. Coconut water is a drink. Coconut milk (endosperm) is used to produce coconut honey, and to make media for growing plant tissues. Coconut oil has special qualities and retains a distinctive advantage for certain uses.

In Florida, coconut palms are grown for their aesthetic qualities. They provide the tropical atmosphere which sets south Florida apart from other areas of the continental United States. The nursery industry in Florida is one of the leading producers of ornamental palms for landscape and interior plantings. However, in the past two decades more than 50% of Florida's 1 million coconut palms and over 80% of Jamaica's 4.3 million coconut palms have been killed by a pandemic disease called lethal yellowing.

The date palm (*Phoenix dactylifera* L.), which is an important plant in the arid tropics and has been associated with the life and culture of the Middle East for more than 5,000 years [3, 4], is also affected by lethal yellowing.

Lethal yellowing (LY) was first reported in the western part of Jamaica [5]. It moved to the eastern coconut growing areas of the island in the 1960s [6]. In a short period, more than 50% of 4.3 million coconut palms in Jamaica alone have been killed by this disease. The annual death rate is estimated at 1 million productive palms.

Lethal yellowing was reported in Key West, Florida in 1965 [7], killing three-fourths of 20,000 coconut palms before disappearing from the island in the late 1960s. In 1971, it appeared in mainland Florida. Since then, the disease has killed approximately 50% of the estimated 1 million coconut palms throughout the coconut-growing areas of Florida. As its impact on the economy and environment was recognized, two independent research teams were organized in Jamaica and Florida to investigate the etiology, epidemiology, ecology, and control of lethal yellowing. This report summarizes various researches on lethal yellowing and discusses the potential for developing methods of disease management.

GEOGRAPHICAL DISTRIBUTION

Information about the geographical distribution of lethal yellowing is mostly based on symptomatology, except in the United States, Mexico, Jamaica, and Togo, where the field diagnosis has been confirmed by electron microscope examinations and the remission of symptoms after antibiotic treatments. Lethal yellowing has been reported in the Caribbean, including the Bahamas [7], Cayman Islands [8], Cuba [9, 10], Dominican Republic [11], Haiti [12], and Jamaica [5], and in North and Central America, including the United States (Florida and Texas) and Mexico [13] and Honduras [14]. In West Africa, this disease is known as Cape St. Paul wilt in Ghana [15, 16] and as Kaincopé in Togo [16, 17, 18], as Kribi in Cameroon [19], Dahomey [16] and as Awka in Nigeria [13].

DISEASE SYNDROME

Symptomatology and Disease Development

The lethal yellowing symptoms in immature coconut palms start with yellowish discoloration from the distal ends of leaf fronds, and are terminated by a final stage of decay of the spear leaves and palm hearts (Plate 5.1). The symptoms in fruit-bearing palms are premature nut fall in all stages of development (Plate 5.2), blackening and necrosis of inflorescences in opened and unopened spathes, and root rot (Plates 5.3, 5.4). A yellow flagleaf may also appear in the middle of the crown at this time. Later, as the disease progresses, the yellow discoloration moves upwards into the crown followed by the decay of the spear leaves and hearts, and the crown topples from the tree leaving a pole-like naked trunk (Plates 5.5, 5.6). The syndrome is complete in three to five months. Similar symptoms appear in *Corypha elata, Pritchardia* spp., *Arikuryroba schizophylla, Trachycarpus fortunei, Dictyosperma album,* and *Hyophorbe verschaffeltii,* but other palms, such as *Veitchia merrillii, Chrysalidocarpus cabadae, Borassus flabellifer, Caryota mitis,* and *Phoenix* spp., do not exhibit leaf yellowing. Instead the leaves turn brown, with water-soaked marks appearing along the pinnae. As the disease progresses, browning extends to the rest of the fronds. In *Veitchia* palms, the browning leaves often break at the joint of the midrib and the leaf base. The spear leaves remain on top of the dead trees (Plates 5.7, 5.8). In *Pritchardia* palms, the death of the spear leaf always occurs first.

Physiological Effects

Little is known about the physiology of lethal yellowing. The study on the water relations of affected palms has revealed the absence of diurnal fluctuation in water potential. This phenomenon could be detected as early as two weeks before the initial visual symptom development [20]. Other studies indicated that water transport in the affected palms was absent or drastically reduced. When palms were injected with ^{32}P solution in the trunk, peak radioactive ^{32}P counts were obtained in healthy leaves after four hours, whereas no activity was measured in diseased leaves even after five days. Similar results were obtained when ^{32}P was applied through palm roots [13]. Another study using silicone stomatal impression technique indicated that stomatal closure occurred in the diseased leaves; no such phenomenon was observed in the healthy palms [13].

The study of electrophoretic analysis of proteins from samples of petiole bases of infected *Veitchia* palms has confirmed a unique protein band at Rf 0.78 ± 0.1. The intensity of this band is proportional to the severity of symptoms. No such band was detected in the healthy samples. Unfortunately, this detection technique did not work with infected coconut palms [13].

Thin layer chromatography has demonstrated that free arginine was readily detected in leaf samples from palms susceptible to lethal yellowing and the intensity of arginine spots on the chromatograms was apparently related to the degree of susceptibility of the palm species [13]. However, subsequent study using HPLC failed to confirm this relationship (T.K. Broschat, personal communication).

HOST RANGE

Lethal yellowing has a wide range of hosts: at least 30 palm species in 7 of the 15 major taxonomic groups [13], and possibly one non-palm host, *Pandanus utilis* [21a]. The susceptibility of palm species to lethal yellowing varies greatly. Palms highly susceptible to lethal yellowing include *Arenga engleri* Becc., *Cocos nucifera* L., *Corypha elata* Roxb., *Phoenix dactylifera* L., *Pritchardia affinis* Becc., *P. pacifica* Seem. & H. Wendl., *P. thurstonii* F.J. Muell. & Drude, and other *Pritchardia* spp. The moderately susceptible species include *Arikuryroba schizophylla* (Mart.) L.H. Bailey, *Borassus flabellifer* L., *Caryota mitis* Lour., *Dictyosperma album* (Bory) H. Wendl. & Drude ex Scheffer, *Hyophorbe verschaffeltii* H. Wendl., *Latania* spp., *Phoenix canariensis* Hort. ex Chabaud., *Trachycarpus fortunei* (Hook.) H. Wendl., and *Veitchia merrillii* (Becc.) H.E. Moore.

Three species are known to be slightly susceptible to lethal yellowing: *Chrysalidocarpus cabadae* H.E. Moore, *Phoenix reclinata* Jacq., and *Veitchia* spp.

The following 10 palm species are known to be susceptible to lethal yellowing, but their numbers are too few to determine their relative susceptibility: *Aiphanes lindeniana* (H. Wendl.) H. Wendl., *Allagoptera arenaria* (Gomes) O. Kuntze, *Gaussia attenuata* (O.F. Cook) Becc., *Howea belmoreana* (C. Moore & F.J. Muell.) Becc., *Livistona chinensis* (Jacq.) R. Br. ex Mart. *Nannorrhops ritchiana* (W. Griffith) J.E.T. Aitch, *Neodypsis decaryi* Jumelle, *Phoenix sylvestris* (L.) Roxb., *P. remota* Becc., and *Ravenea hildebrandtii* H. Wendl. ex Bouche [13, 22].

EPIDEMIOLOGY

Lethal yellowing is not only a fast-killing disease but also a rapidly spreading disease. In Dade County, Florida 75% of the coconut population was lost to lethal yellowing in three years. In general, the spread is of two types. The initial spread involves a local center with one or two infected trees. Later, new infections randomly appear surrounding the initial infection site. The first eight months of the lethal yellowing epidemic in Miami in 1971 was basically demonstrated in primary focus development. The second type of spread is a jump spread followed by the pattern of local spread. Since 1973, the spread has been erratic within the lower east coast and southern portion of Martin County, Florida. Nearly 100,000 coconut palms and several thousands of other palms have been killed in such a way in a 10-year period.

In Date County, Florida where the occurrence of nearly 20,000 infected trees and their locations were accurately recorded, McCoy was able to calculate the apparent infection rate of lethal yellowing based on proportional analysis. The apparent rate of spread was lower on shoreline sites than on inland sites [23]. Similar results have been obtained by Steiner for Kaincopé disease in West Africa [24]. Britt calculated lethal yellowing dispersal gradients based on spatial analysis of disease incidence data for an area of 200 square miles [25].

ETIOLOGY

Since the early 1900s, a number of researchers have investigated the cause of lethal yellowing. Initial emphasis was placed on fungi or bacteria as possible causal agents [7, 12, 26, 27, 28]. Nematodes were implicated [7, 29]. Abiotic factors such as nutrient deficiency, salt damage, toxic materials, and soil conditions were also suspected causes [12, 30, 31, 32]. A virus was strongly considered to be the causal agent of lethal yellowing as no other disease-causing agent could be found nor did applications of fertilizers, fungicides, bactericides, and pesticides control the disease [6, 33, 34, 35, 36, 37]. Unsuccessful attempts to transmit the causal agent by dodder and mechanical inoculation have also been reported [33, 38].

In 1972, the first electron micrographs of ultrathin sections from diseased inflorescences were published revealing the presence of mycoplasma-like organisms (MLOs) and they were found in diseased but not in healthy coconut palms. No virus-like particles or microorganisms other than MLO were detected in the ultrathin sections of the phloem, xylem, or parenchyma tissue [39]. A note published shortly thereafter by Breakbane et al. (1972) confirmed the presence of MLO in diseased coconut palm material from Jamaica [40]. Subsequently, other researchers also observed MLO in association with lethal yellowing in Jamaica [41], Florida [42], Togo [43, 44], and Cameroon [19]. Further evidence supporting MLO etiology of the disease came from the results of tetracycline treatments [18, 45, 46].

SEARCH FOR LETHAL YELLOWING VECTORS

Many species of suspected vectors have been tested by researchers in the Caribbean and Florida since 1940. The test species included members of Cicadellidae, Fulgoridae, Derbidae, Delphacidae, Issidae, Aphidae, Coccoidea, Pseudococcidae, Aleyrodidae, Cixiidae, Cicadidae, Tingidae (Hemiptera), and Thysanoptera [6, 7, 26, 33, 47, 48, 49, 50, 51]. These unsuccessful studies were summarized in recent reviews [52, 53]. Other suspected vectors included eriophyid mites [54] and nematodes [7]. The mycoplasma etiology and epidemiological evidence of lethal yellowing has led the search for potential vectors among the leafhoppers (Cicadelloidea) and the planthoppers (Fulgoroidea), as the other known vectors of MLO are Auchenorrhynchus insects [55]. Heinze and Schuiling (1970) were apparently successful in transmitting the lethal yellowing agent to caged palms by using mixtures of *Myndus crudus* Van Duzee and other Auchenorrhynchus insects, but because of an extremely low rate of transmission their results were discounted [54]. Recent studies have strongly implicated *M. crudus* as the vector, based on the symptom development of caged palms after the introduction of wild populations of this insect, and MLOs were found in the infected tissue [13, 56]. However, numerous attempts have failed to prove that *M. crudus* (Plate 5.9) reared from either colonies or wild populations could acquire and transmit the lethal yellowing agent from diseased palms to healthy palms (J.H. Tsai, unpublished) [21b, 53, 57].

CONTROL AND MANAGEMENT

No single measure has been proved to effectively combat lethal yellowing. An integrated management program is currently recommended which includes eradication, quarantine, chemotherapy, the development of resistant palms, and vector control. Eradication of infected palms can be effective in slowing the lethal yellowing epidemic, especially at the early stage of disease spread. Quarantine measures, which have been imposed by several states in the United States, prevent the introduction of the lethal yellowing agent or its vector into unaffected areas. Although antibiotic treatment with oxytetracycline cannot cure the disease, it can prolong the lives of infected palms and may maintain susceptible palms while a transition is made from susceptible to resistant palms [13, 46]. Two resistant varieties of coconut palm, the Malayan dwarf and a hybrid, Maypan, have been developed and released by the Coconut Industry Board in Jamaica and Lethal Yellowing Research Teams from the FAO, UN and the ODM, UK. They are highly resistant to lethal yellowing in both Jamaica and Florida. In the last few years, about one million coconut palms of these two varieties have been planted in Jamaica [58]. Certain insecticides may slow the spread of the disease by killing the insect vector [13], but their use in a large area may prove to be economically and environmentally unfeasible.

REFERENCES

1. Oliver, P.L. *The Pacific Islands.* Garden City, New York: Doubleday and Co., p. 456, 1961.

2. Moore, H.E., Jr. Palms in the tropical forest ecosystem of Africa and South America. In: *Tropical Forest Ecosystems in Africa and South America: A Comparative Review.* B.J. Meggers, E.S. Ayensu and W.D. Duckworth, eds. Washington, DC: Smithsonian Institute Press, p. 63, 1978.

3. Corner, E.J.H. The natural history of palms. London: Weidenfeld and Nicholson, p. 393, 1966.

4. Goor, A. The history of the date through the ages in the Holy Land. *Econ. Bot.* 21, 33, 1967.

5. Fawcett, W. Report on the coconut disease at Montego Bay. *Bull. Bot. Dept. Jamaica,* 23, 2, 1891.

6. Grylls, N.E., and P. Hunt. Studies on the aetiology of coconut lethal yellowing in Jamaica, by mechanical and bacterial inoculations and by insect vectors. *Oleagineux,* 26, 543, 1971.

7. Martinez, A.P. Lethal yellowing of coconut in Florida. *FAO Plant Prot. Bull.*, 13, 25, 1965.

8. Seal, J.L. Coconut bud rot in Florida. Univ. of Florida, Florida Agric. Sta. Bull. No. 199. p. 27, 1928.

9. De la Torre, C. La enfermedad de los cocoteros. *Rev. de la Faculdad de Let. y Cienc.* (Havana), 2, 269, 1906.

10. Mijailova, P.T. Informe sobre la investigacion de la enfermedad 'Pudricion del cogollo del cocotero'. *Revista Agric. Habana*, 1, 74, 1967.

11. Schieber, E., and E. Hichez-Frias. Lethal yellowing disease of coconut palms in the Dominican Republic. *Phytopathology,* 60, 1542, 1970.

12. Leach, R. The unknown disease of coconut palm in Jamaica. *Trop. Agric.* (Trinidad), 23, 50, 1946.

13. McCoy, R.E., F.W. Howard, J.H. Tsai, H.M. Donselman, D.L. Thomas, H.G. Basham, R.A. Atilano, F.M. Eskafi, L. Britt, and M.E. Collins. Lethal yellowing of palms. Univ. of Florida, IFAS Agric. Exp. Sta. Tech. Bull., No. 834, p. 100, 1983.

14. Anonymous. Disease still killing Mexico's palm trees. *National Geographic,* 176, 1989.

15. Leather, R.I. Further investigations into the Cape St. Paul wilt of coconuts of Keta, Ghana. *Emp. J. Expt. Agric.*, 27, 67, 1959.

16. Maramorosch, K. A survey of coconut diseases of uncertain etiology. Rome: FAO, p. 39, 1964.

17. Ollagnier, M., and G. Weststeijn. Les maladies du cocotier auxiles Caraibes: Comparison avec la maladie de Kaincopé au Togo. *Oleagineux,* 16, 729, 1961.

18. Steiner, K.G. Remission of symptoms following tetracycline treatment of coconut palms affected with Kaincopé disease. *Plant Dis. Rep.*, 60, 617, 1976.

19. Dollet, M., J. Giannotti, J.L. Renard and S.K. Ghosh. Etude d'un jaunissement letal des cocotiers au Cameroun: la maladie de Kribi. Observations d'organismes de type mycoplasmes. *Oleagineux,* 32, 317, 1977.

20. McDonough, J., and M.H. Zimmerman. Effect of lethal yellowing on xylem pressure in coconut palms. *Principes,* 23, 132, 1979.

21a. Thomas, D.L., and H.M. Donselman. Mycoplasma-like bodies and phloem degeneration associated with declining Pandanus in Florida. *Plant Dis. Rep.*, 63, 911, 1979.

21b. Tsai, J.H. Attempt to transmit lethal yellowing of coconut palms by the planthopper, *Haplaxius crudus. Plant Dis. Rep.*, 61, 304, 1977.

22. Maramorosch, K., and P. Hunt. Lethal yellowing disease of coconut and other palms. In: *Mycoplasma Diseases of Trees and Shrubs.* K. Maramorosch and S.P. Raychaudhuri, eds. New York: Academic Press, p. 185, 1981.

23. McCoy, R.E. Comparative epidemiology of the lethal yellowing, Kaincopé, and Cadang-cadang diseases of coconut palm. *Plant Dis. Rep.*, 60, 498, 1976.
24. Steiner, K.G. Epidemiology of Kaincopé disease of coconut palms in Togo. *Plant Dis. Rep.*, 60, 613, 1976.
25. Britt, L.L. A study in spatial diffusion: lethal yellowing in *Cocos nucifera*. MSc thesis, Univ. of Miami, 115 pp., 1981.
26. Grylls, N.C., and N.A. Bor. Investigations on the etiology of lethal yellowing of coconuts in Jamaica. 2. The role of insects as possible vectors. FAO Technical Working Party on Coconut Production, Protection, and Processing, 3rd session, Indonesia, 1968.
27. Grylls, N.E. P. Hunt and N.A. Bor. An approach to the study of the aetiology of coconut lethal yellowing disease. Paper presented at the First International Congress of Plant Pathology, London, July 1968.
28. Martyn, E.B. 1945. Coconut diseases in Jamaica (I) 'bronze leaf wilt' and other diseases affecting the bud of coconuts. *Trop. Agric.* (Trinidad), 22, 51, 1945.
29. Latta, R.K. Attempts to relate nematodes to lethal yellowing disease of coconuts in Jamaica. *Trop. Agric.* (Trinidad), 43, 59, 1966.
30. Chen. R.A. Nutritional aspects of lethal yellowing in coconuts. *Trop. Agric.* (Trinidad), 43, 211, 1966.
31. Fiskell, J.G.A., A.P. Martinez and L.G. Van Weerdt. Chemical studies on the roots and leaves of coconut palms affected by lethal yellowing. *Proc. Florida State Hort. Soc.* p. 408, 1959.
32. Innes, R.F. The manganese content of leaf and inflorescence tissue in relation to the 'unknown disease' of the coconut palm in Jamaica. *Trop. Agric.* (Trinidad), 26, 57, 1949.
33. Carter, W. Susceptibility of coconut palm to lethal yellowing disease. *Nature* (London), 212 (5059), 320, 1966.
34. Corbett, M.K. Disease of the coconut palm. *Principes* 3, 5, 1959.
35. Martinez, A.P., and D.A. Roberts. Lethal yellowing of coconuts in Florida. *Proc. Florida State Hort. Soc.*, 80, 432, 1967.
36. Nutman, F.J., and P.M. Roberts. Lethal yellowing: The 'unknown disease' of coconut palms in Jamaica. *Emp. J. Exp. Agric.*, 23, 257, 1955.
37. Price, W.C., A.P. Martinez and D.A. Roberts. Reproduction of the coconut lethal yellowing syndrome by mechanical inoculation of young seedlings. *Phytopathology*, 58, 593, 1968.
38. Grant, T.J. Lethal yellowing disease of coconut. FAO Rep. No. TA2367 to the Government of Jamaica, 1967.
39. Plavsic-Banjac, B., P. Hunt and K. Maramorosch. Mycoplasma-like bodies associated with lethal yellowing disease of coconut palms. *Phytopathology*, 62, 298, 1972.
40. Breakbane, A.B., C.W. Slater and A.F. Posnette. Mycoplasmas in the phloem of coconut, *Cocos nucifer* L., with lethal yellowing disease. *J. Hort. Sci.*, 47, 265, 1972.
41. Heinze, K.G., H. Petzold and R. Marwitz. Beitrag zur Atiologie der todlichen Vergilbung der Kokospalme. *Phytopathol. Z.*, 74, 230, 1972.
42. Parthasarathy, M.V. Mycoplasma-like organisms in the phloem of palms in Florida affected by lethal yellowing. *Plant Dis. Rep.*, 57, 861, 1973.
43. Dollet, M., and J. Giannotti. Maladie de Kaincopé: Presence de mycoplasmes dans le phloeme des cocotiers malades. *Oleagineux*, 31, 169, 1976.
44. Neinhaus, F., and K.G. Steiner. Mycoplasma-like organisms associated with Kaincopé disease of coconut palms in Togo. *Plant Dis. Rep.*, 60, 1000, 1976.
45. Hunt, P., A.J. Dabek and M. Schuiling. Remission of symptoms following tetracycline treatment of lethal yellowing infected coconut palms. *Phytopathology*, 64, 307, 1974.
46. McCoy, R.E. Remission of lethal yellowing in coconut palm treated with tetracycline antibiotics. *Plant Dis. Rep.*, 56, 1019, 1972.

47. Dabek, A.J. Current attempts to transmit lethal yellowing disease in Jamaica. *Proc. 3rd Mtg. Int'l. Council on Lethal Yellowing*, Univ. of Florida. FL-78-2. p. 22, 1978.

48. Eden-Green, S.J. Rearing and transmission techniques for *Haplaxius* sp., a suspected vector of lethal yellowing disease of coconut palms in Jamaica. *Ann. Appl. Biol.*, 89, 173, 1978.

49. Eden-Green, S.J. and M. Schuiling. Root acquisition transmission tests with *Haplaxius crudus* (Homoptera: Cixiidae) and *Proarna hilaris* (Homoptera: Cicadidae). *Principes*, 20, 66, 1976.

50. Heinze, K.G. Lethal yellowing disease of coconut. FAO Rep. No. TA3152 to the Government of Jamaica, 1972.

51. Johnson, C.G. A review of the vector experiments in Jamaica from 1962 to 1971. *Principes*, 20, 65, 1976.

52. Johnson, C.G., and S.J. Eden-Green. The search for a vector of lethal yellowing of coconut in Jamaica: Reappraisal of experiments from 1962–1971. *FAO Plant Prot. Bull.*, 26, 137, 1978.

53. Tsai, J.H. Lethal yellowing of coconut palm: Search for a vector. In: *Vectors of Plant Pathogens*. K.F. Harris and K. Maramorosch, eds. New York: Academic Press, p. 177, 1980.

54. Heinze, G.G., and M. Schuiling. Tenth Rep. Res. Dept. Coconut Industy Board, Jamaica, p. 84, 1970.

55. Tsai, J.H. Vector transmission of mycoplasmal agents of plant diseases. In: *The Mycoplasmas*. M.F. Barile, S. Razin, J.G. Tully and R.F. Whitcomb, eds. New York: Academic Press, p. 265, 1979.

56. Howard, F.W., and D.L. Thomas. Transmission of palm lethal decline to *Veitchia merrillii* by planthopper, *Myndus crudus*. *J. Econ. Ent.*, 73, 715, 1980.

57. Tsai, J.H., and D.L. Thomas. Transmission of lethal yellowing mycoplasma by *Myndus crudus*. In: *Mycoplasma Diseases of Trees and Shrubs*. K. Maramorosch and S.P. Raychaudhuri, eds. New Delhi: Academic Press, p. 211, 1981.

58. Romney, D.H., and H.C. Harries. Distribution and impact of lethal yellowing in the Caribbean (Abstr.) Proc. 3rd Meeting Int. Counc. Lethal Yellowing. Univ. Florida Agric. Res. Cent. Ft. Lauderdale, Publ. FL-78-2, p. 6, 1978.

59. Harrison, N.A., C.M. Bourne, R.L. Cox, J.H. Tsai and P.A. Richardson. DNA probes for detection of mycoplasmalike organisms associated with lethal yellowing disease of palms in Florida. Phytopathology 81, (in press), 1991.

Addendum

Lethal yellowing and LY-like diseases presumed to be caused by MLOs pose the most important threat to global coconut and date productions. The epidemic of LY in Mexico and Honduras may serve as a springboard for the disease to spread to the entire Central America and Pacific region. The report of MLO association with coconut root wilt disease in India makes the issue more complex. All these diseases in three continents show many similarities and dissimilarities in symptomatology, varietal susceptibility and epidemiology. The identity and relationship between these MLOs is of fundamental importance to entomologists, plant pathologists, epidemiologists, plant breeders and those concerned with plant health in general. Research in these areas have been hindered for the last two decades due to our inability to culture the MLOs *in vitro*.

Because of recent advances in molecular biology and immunology, we have been able to adopt the techniques and developed DNA probes for detection of LY MLOs in Florida [59]. The DNA-based diagnostic assays should greatly enhance our future research including the identification of MLOs associated with LY complex, vector and epidemiological studies and plant breeding and quarantine.

Chapter 6

STRATEGIES USED IN ETIOLOGICAL RESEARCH ON COCONUT AND OIL PALM DISEASES OF UNKNOWN ORIGIN

Michel Dollet

Michel Dollet is Director of the Virology Division at IRHO (Institut de Recherches pour les Huiles et Oléagineux), the Oil Crops Department of CIRAD (Centre de Coopération Internationale en Recherche Agronomique pour le Développement) in Montpellier, France. He is also Chairman of the CIRAD-ORSTOM Laboratoire de Phytovirologie des Régions Chaudes (LPRC). He obtained a PhD from the Université Louis Pasteur in Strasbourg in 1978 and a DSc from the Université des Sciences et Techniques du Languedoc in Montpellier in 1985. He is a member of the Société Française de Phytopathologie, the Association of Applied Biologists, the American Phytopathological Society, the International Organization of Mycoplasmologists and the Society of Protozoologists, and was co-Chairman of the International Council on Lethal Yellowing in 1979. He is currently working on diseases of unknown origin presumed to be due to viruses or viroids that affect oil palm and coconut and viral diseases of groundnut. Since 1980, he has been working on plant trypanosomatids (Phytomonas) in vitro culture, characterization, and transmission by bugs. Address: Division Virologie IRHO. LPRC, CIRADORSTOM, BP 5035, 34032 Montpellier Cedex 1, France.

CONTENTS

INTRODUCTION

In the early 1970s, several diseases of oil palm (*Elaeis guineensis* Jacq.) and coconut (*Cocos nucifera* L.) were considered diseases of unknown origin. Etiological research conducted on these diseases had not succeeded in specifically associating any of the fungi, bacteria, nematodes, or insects with the pathological syndromes studied. Virological research had also been unable to provide conclusive results.

The very nature of the two palms in question largely accounts for the failure of virological research. In fact, there are many plausible explanations for the difficulties encountered in solving these etiological enigmas.

Perennial, Monocotyledonous Plants

Oil palm and coconut can be said to combine all the drawbacks of perennial plants with those of monocotyledonous plants. For example, grafting, a basic technique in phytovirological studies, is impracticable. Transmission trials using implants have usually produced negative results. Oil palm and coconut are planted to be exploited for 20 to 50 years or more; when adult, their stem diameter varies from 20 to 60 cm and their height from 12 to 20 m or more.

Difficulties with Transmission by Mechanical Inoculation

Mechanical inoculation, the other basic virology technique, has never been practicable either. There are undoubtedly many reasons for this failure. However, the existence of numerous inhibitors (primarily tannins) in the tissues is probably one of the major causes.

Tissues

Palms are very fibrous. Fibers are found not only in the stem, but also in the leaf rachis, leaflets, and inflorescences. These fibers, in conjunction with tannin cells and the thick leaflet cuticle, pose problems for histological and ultrastructural studies. The hardness of the tissues and their low water content make it difficult to grind samples for obtaining extracts suitable for purification of any viral particles present.

Incubation Periods and Space Required

The incubation periods for these diseases of unknown origin are often very long, extending from 6 to 18 months or longer, and very large glasshouses are required to house inoculated palms for one or two years.

Since oil palm leaves measure more than one meter long at the age of one year, many plants have to be inoculated to obtain appreciable mechanical transmission percentages; it is easy, therefore, to see why such trials are rarely conducted in glasshouses. If field trials are to be conducted, the long incubation periods involved mean that the number of plants inoculated has to be even greater, so that strict statistical designs may be applied to allow for natural contamination, an operation which always proves to be extremely tricky.

Growth Period

Palms grow slowly. Oil palm germination can be forced in an artificial environment for four to eight weeks at constant temperature and humidity. It takes a further two months to obtain a seedling with two leaves, thereafter a new leaf appears each month. Depending on the variety and on climatological conditions, the first fruits appear only after four to seven years. A considerable amount of time needs to be spent for any kind of study on palms.

Accessibility

Palms are not the most easily accessible plants (not to mention their geographical remoteness). Oil palm has a double row of sharp spines on the leaf petiole, and also on female flower spikes. Sampling and detailed examination of a tree whose leaf crown is more than 10 m above the ground pose certain problems. It is therefore often necessary to fell the tree and, as a result, it is impossible to monitor disease development or to organize sampling schedules.

Dissociating Geographical Locations, Laboratory Plantations

The impossibility of grafting and the near-impossibility of physical transmission deprive virologists of the means of sustaining the disease in glasshouses close to the laboratory. Apart from a few exceptions, plantations are often isolated, far from any well-equipped centres with staff capable of carrying out such research. Many oil palm- and coconut-growing countries in the tropics do not yet have the resources required for studies of this nature. Virologists therefore have to work on single samples mailed to them by collaborators or those collected by themselves on surveys, often repeating the journey or the dispatches until their research is concluded. It would be of course unthinkable and rather impossible to set up a virology laboratory at every site where diseases of unknown origin are observed.

Evolution of Concepts

Besides the many drawbacks associated with the nature of the plant itself, the relatively slow evolution of concepts must also be considered, as with every other disease of unknown origin unlike, in the majority of cases, with more easily approachable systems. Until 1967 and the detection of plant mycoplasmas [1], it had been difficult to diagnose this type of etiological agent in oil palm or coconut. The same was the case of the viroid concept, which, to realize, took more than 20 years of virological research on a much more easily accessible plant than palms, the potato [2].

The evolution in phytopathological concepts, in conjunction with improved laboratory techniques and the adoption of certain strategies arising from the above difficulties, has enabled us to take a considerable step forward in the past 15 years in the knowledge of diseases of unknown origin in oil palm and coconut. These strategies can be illustrated with the help of a few significant examples as follows:

1) Ultrastructural study of well-defined samples by electron microscopy.

2) Comparative study of nucleic acids in diseased and healthy plants.

3) Vector identification by exposing healthy plants to large numbers of insects in cages.

ULTRASTRUCTURAL STUDY BY ELECTRON MICROSCOPY

As it is not possible to use grafts, dodders, or mechanical inoculation, attempts can be made to visualize the pathogen directly in the diseased plant. If successful, the first of Koch's postulates will also be satisfied—'the specific association of a pathogen with a given disease'— but ultrathin tissue sections are yet to be examined by electron microscope. It is difficult to fix and embed oil palm and coconut stem, leaf rachis, or primary root tissues for electron microscope studies. Only three types of tissue can easily be studied using the electron microscope: young inflorescences, very young leaves, and root apices. However, conventional methods will have to be adapted to this material. Hence, embedding resins for electron microscopy will have to be of very low viscosity, such as Spurr's resin, and with conventional epoxy resins (Epon), soaking for long periods in gradual impregnation baths will be necessary [3].

The last but not the least important factor for success is fixing the right type of sample at the right moment. In fact, to the untrained eye, many oil palm or coconut diseases look alike, particularly during their terminal stages, when there are only a few young, more or less yellowed leaves left and the meristem area has consequently begun to rot. Many diseases have therefore been identified as lethal yellowing, fatal yellowing, bud rot, or spear rot. However, experience has often shown that the pathogen can most often be visualized in the early stages of the disease, when symptoms are not very advanced. A thorough general knowledge of the plant and of symptomatology is therefore a vital asset.

Lethal Yellowing

Coconut lethal yellowing has been rife in the Caribbean for more than 150 years. The disease spread to Florida in the 1950s and appeared in Mexico in 1982 [4].

In West Africa, pathological syndromes identical to lethal yellowing have been observed in various countries, each of which has given them a local name: Cape St. Paul wilt in Ghana, Kaincopé disease in Togo, Awka disease in Nigeria, and Kribi disease in Cameroon. The diseases have been observed for almost 60 years. In East Africa, an identical disease was described for the first time in 1972 in Mozambique [5]. It is probably the same disease as the one known as lethal disease in Tanzania.

In each case, the disease begins with nut fall, irrespective of the size and maturity of the nuts. This is followed by the browning and necrosis of recently opened inflorescences, accompanied or followed by the first signs of lower leaf yellowing (Plate 6.1).

It was only after decades of research that the etiology of these diseases was defined, and mycoplasmas specifically associated with lethal yellowing [6, 7], with various yellowing diseases in West Africa [8–10], and then with lethal disease in Tanzania [11] were detected under the electron microscope. The number of articles that have been published on the subject does not, however, truly reflect the difficulties encountered in these studies. For example, the mycoplasma concentration in sieve tubes from trees affected by Kaincopé disease is often very low (between 2 and 10 organisms in a longitudinal section). In general, only 5 to 20% of sieve tubes contain organisms. Lastly, mycoplasma observation is strictly dependent on the type of sample fixed: there are more chances of conducting a mycoplasma diagnosis when using samples taken at peduncle level of unopened inflorescences (inflorescences 8-7-6 starting from the spear = leaf 0), at or before the onset of yellowing [3, 7, 9, 10], or even earlier, during nut fall. Examining samples at a later stage can lead to incorrect diagnoses, due to the rapid appearance of a great many secondary contaminants, including bacteria [12], or even the failure to observe mycoplasmas. In Florida, the bases of very young leaves are currently used to diagnose lethal yellowing [4].

The etiological analysis of mycoplasmas by means of examination under a microscope was confirmed by trials using tetracycline antibiotic treatments. Coconut, by virtue of its many vascular bundles throughout the thickness of the stem, is well suited for injection. Injecting oxytetracycline before yellowing reaches the leaf crown leads to remission of the yellowing, whereas penicillin and streptomycin are ineffective [13–15].

Oil palm Marchitez, Coconut Hartrot

In the early 1970s, oil palm growers in Latin America were confronted with a wilt disease known as Marchitez caso A and Marchitez caso B (wilt case A and wilt case B). In fact, the two diseases were first differentiated in 1976 after a symptomatological study. One was called 'ring spot disease', due to the characteristic symptom observed on leaves, and the name 'Marchitez sorpresiva' was adopted for the other form of wilt, a sudden wilting characterized by very rapid drying out of leaves, from the base upwards (Plate 6.2) [16]. Coconut hartrot has been observed in Surinam for almost a century, and was once called 'lethal yellowing' or 'unknown disease'. In fact, there is a stage in the development of the disease in which the leaves turn yellow—or brown (depending on the variety)—preceded or accompanied by inflorescence necrosis. The only difference from lethal yellowing is the general absence of immature nut fall (Plate 6.3).

In theory, Marchitez and hartrot diseases had nothing in common. Hartrot had been in existence for a long time, and was concentrated mostly in Surinam, while Marchitez had only appeared recently, just after oil palm cultivation developed in Latin America in the beginning of the 1960s, at first in Colombia, and then in Peru and Ecuador.

Two different teams conducted independent studies of the two diseases. In both cases, electron microscopy revealed the presence of flagellate protozoa, *Kinetoplastidae*, of the Trypanosomatidae family, in the sieve tubes of diseased trees [17, 18]. A similar electron microscope analysis was then carried out on a coconut wilt in Trinidad, locally known as Cedros wilt [19].

In all three cases, analysis was carried out on young inflorescences or young leaves. These organs once again proved to be excellent material for the study, but as with lethal yellowing, the chances of observing the pathogen decreased as the disease developed. Hence, for Marchitez, it is easier to carry out the analysis during the first stage of the disease—browning of leaflet and lower leaf tips—or even earlier, at the end of the incubation period. During the final stage of the disease, root examination reveals high bacterial and paramecial contamination and an absence of trypanosomes.

However, there is a significant difference between hartrot and Marchitez. By virtue of their large size, these trypanosomatids can—now that the 'plant trypanosome' concept has been established—be identified using light microscope [20]. With oil palm, the roots are most suitable for examination (percentage of infected roots is usually more than 80%), whereas with coconut, trypanosomatids are most likely to be found in the stalks of inflorescences still protected by their spathes.

The pathogenicity of these organisms has not yet been proved in accordance with Koch's postulates. This is essentially due to the difficulties encountered in trying to grow them *in vitro*. Now that cultures have been established [21] and vectors are known [22], it should be possible to demonstrate their pathogenicity. However, it is easy to presume their pathogenicity, since they are not observed in healthy oil palms or in trees affected by other diseases in Latin America.

COMPARATIVE STUDY OF NUCLEIC ACIDS IN DISEASED AND HEALTHY PLANTS

It is not always possible to detect a pathogen directly, even with the electron microscope. The possibility decreases with the decrease in size of the pathogen and its dilution in the plant. It is obviously easier to visualize mycoplasmas 300 to 500 nm in diameter in ultrathin sections than viral particles 30 to 80 nm in diameter. The smaller the viral particles—about the same size as ribosomes—the more difficult they are to identify, unless they cluster together forming specific particle agglomerations. If they are small and not very concentrated, analysis will be very uncertain, unless particular enzymatic treatments [23, 24] and numerous long and tedious observations are carried out. Lastly, it is impossible to visualize viroids directly, due to their miniature size—246 to 375 nucleotides—and the lack of a protein shell.

This lack of direct pathogen visualization *in situ* can be compensated for by the detection of the pathogen genome—the RNA or DNA of the virus—or the viroid itself, using biochemical and molecular biology techniques. The technique generally consists in extracting the nucleic acids from healthy and diseased plants, comparing their electrophoretic mobilities, and identifying one or more extra bands in samples of the diseased plants.

Cadang-Cadang and Tinangaja Diseases

The first example of using nucleic acids to detect the pathogen of an oil palm or coconut disease was Cadang-cadang disease in the Philippines. Symptoms develop slowly and the disease is revealed by a change in nut shape: nuts become smaller and round, with extensive scarification. However, the most recognizable symptom at this stage is the occurrence of small yellow leaf spots, which appear olivaceous or water-soaked when viewed by reflected light (Plate 6.4). New inflorescences are increasingly smaller and become sterile. The leaf spots eventually coalesce, the leaves look as if chlorosed and become increasingly shrunken.

To explain the etiology of the disease, a wide range of hypotheses were considered, from nutrition deficiencies to viruses, including volcanic eruptions and nematodes [25]. In 1973, the search for possible virus particles in the Philippines was restricted by a lack of equipment, particularly ultracentrifuges. Randles undertook to concentrate possible virus particles in the Philippines by producing a precipitate using 5% polyethylene glycol (PEG 6000). A precipitate obtained by low-speed centrifugation was used in Australia to visualize the virus particles concentrated in this way with an electron microscope [26, 27]. No defined particles could be associated with the disease. It was then decided to extract the nucleic acids from the PEG precipitates, in the hope of detecting the nucleic acid of the virus causing the disease. The nucleic acids were extracted using phenol in the presence of sodium dodecyl sulfate (SDS). After enzyme digestion using *Streptomyces griseus* protease, a second phenol extraction was conducted, finally precipitating the nucleic acids by ethanol. The nucleic acids extracted in this way were separated by electrophoresis on 2.5% polyacrylamide gel in Tris-borate-EDTA buffer. This revealed two bands sensitive to RNase, specifically associated with the disease: ccRNA1 and ccRNA2, the two monomer and dimer forms of the Cadang-cadang viroid (CCCV) [28, 29]. The pathogenicities of these RNAs were then demonstrated by pressure injection in young, healthy trees [30, 31].

A similar technique made it possible six years later to reveal the etiological role of a viroid in the coconut disease known as Tinangaja disease on Guam [32]. The viroid exists in two bands—monomer and dimer—as does CCCV, but having only 64% sequence homology [33].

However, when the protocol for CCCV extraction in coconut was strictly followed, it was difficult to detect CCCV in other palms, particularly *Elaeis guineensis* Jacq. (oil palm) and *Corypha elata* Roxb. (buri palm), both affected by Cadang-cadang disease in the Philippines. It was necessary to modify the protocol by adding 2% polyvinyl pyrrolidone (insoluble form) and clarifying with chloroform before precipitation using PEG [34].

It should be stressed that for these two viroids—CCCV and Tinangaja—macromolecule precipitation with PEG was used, which was initially intended for precipitation of virus particles, unlike in the case of the viroids of other plants.

Coconut Foliar Decay

All the new coconut varieties introduced in Vanuatu (formerly New Hebrides) since the 1960s are susceptible to a lethal disease known as coconut foliar decay (CFD). Only the 'local' variety, Vanuatu tall, planted at the turn of the century, shows no symptoms of the disease.

The first symptom of CFD is leaf yellowing in the middle of the crown (leaves 7 to 11 starting from the unopened spear). The leaves die prematurely and hang down from the stem between the lower leaves. Yellowing gradually spreads to the youngest leaves, while the lower leaves also gradually die to senility (Plate 6.5). The tree dies within two years after the initial symptoms appear. The existence of this generalized coconut yellowing led certain authors to attribute it to lethal yellowing. However, there are several distinguishing factors between the two diseases. In the case of CFD:

(1)   There is no immature nut fall.
(2)   Yellowing does not start with the lower leaves.
(3)   Young inflorescences do not turn brown or become necrotic.
(4)   Close examination of yellowed leaflets shows the existence of brown spots (Plate 6.6) which do not exist in lethal yellowing.

These differences, based on symptomatology, are confirmed by the absence of mycoplasmas in diseased tissue [35] and the lack of response to antibiotic treatments [36]. The hypothesis of a similar etiology as of Cadang-cadang or Tinangaja disease was also rejected after a comparative study of nucleic acids by electrophoresis and molecular hybridization techniques [3, 35].

However, electrophoresis in 3.3% polyacrylamide 8 M urea gel revealed, after silver staining, the existence of a RNase-resistant, DNase I-sensitive band, specifically associated with CFD [37]. A study of this DNA band by electron microscopy, and its behavior in two-dimensional electrophoresis, its resistance to end-labeling following treatment with alkaline phosphatase and polynucleotide kinase, showed that it was a single-stranded circular DNA of about 1300 nucleotides in size [38]. The fact that the disease is transmitted by a planthopper [39] and that a DNA is specifically associated with it favors a viral etiology hypothesis. Twenty nm icosahedral particles which copurify with CFD-DNA have recently been detected [40]. A virus of this size with such a small circular

DNA does not fit into any plant virus group. It remains to be proved whether these particles do in fact cause CFD. It should be noted that as with Cadang-cadang, both CFD-DNA isolation and virus particle isolation involved precipitation with PEG 6000. However, due to the very low concentration of both CFD-DNA and viral particles, it was necessary to introduce other technical concepts in this etiological research: silver staining of polyacrylamide gels for DNA analysis; absorption with an antiserum specific to healthy coconut palm antigens; and isopycnic nycodenz gradients, for particle purification.

Oil Palm Bud Rot

In the Amazon region of Ecuador, the oil palm plantations set up in the 1970s are all affected by a decay known locally as bud rot. Similar decays, known as fatal yellowing, lethal spear rot, etc., exist in other regions of northern Latin America [41]. It is impossible to say whether bud rot is the same as Turbo disease (another disease of unknown origin), which destroyed the plantation of the same name in northern Colombia in the years 1967–1971 [41]. There is also a very similar oil palm disease in Para State, Brazil and in Surinam, originally called 'spear rot' and then—in Brazil—'fatal yellowing'. More recently, high percentages of a disease of the bud rot type were seen in 1989 in various Colombian plantations, notably in Llanos.

In fact, diseased trees can be identified by a slight yellowing of young leaves (leaves 1, 2 and 3) (Plate 6.7). At this stage, spear rot is already noticeable (Plate 6.8). This rot seems to spread 'externally' (by contact), little by little to younger and younger spears (simply by the gravitational flow of the substance resulting from the decomposition of rotting tissues higher up) until it reaches the tree's meristem area, leading to its death (Plate 6.9). The lower leaves do not show any symptoms at any stage of the disease; however, specking and vein banding are sometimes visible on leaves 4 to 9 while the tips of leaves 4 to 7 are matt bronze in color (Plate 6.10). [42].

As ultrastructural studies under the electron microscope failed to detect a specific pathogen, etiological research was steered towards comparing nucleic acids from diseased and healthy plants [42]. The extraction technique used was based on that for the Cadang-cadang viroid, with precipitation using PEG 6000 (8% by volume) and extraction with phenol. The nucleic acids precipitated using ethanol were finally treated with cetyl trimethyl ammonium bromide to remove excess polysaccharides. Electrophoresis in 3.3 to 5% polyacrylamide gel did not reveal any particular bands.

Schumacher et al. [43] developed another electrophoresis technique—two-dimensional or bi-directional gels—to detect viroids and virusoids. This technique, based on the structural properties of virusoids and viroids—covalently closed and base-paired single-stranded RNA—includes preliminary migration under native conditions, then a second migration of the zone thought to contain the hypothetical viroid under partly denaturing conditions. In the second direction, the migration of the viroids, in the form of circular molecules, is slower. This technique revealed a band of lesser mobility which migrates at the same level as the chrysanthemum stunt viroid, in oil palms affected by bud rot in Ecuador and also in supposedly healthy trees in the same area [42]. Using the same type of technique but carrying out both migrations on the same polyacrylamide gel (return electrophoresis), Singh et al. revealed a similar molecule in oil palms affected by fatal yellowing in Brazil and in non-symptomatic trees in the same plantation [44].

This band, sensitive to pancreatic RNase, soluble in 2 M lithium chloride, was also detected in healthy oil palms in Africa, where there is currently no bud rot or fatal yellowing problem [45]. At that point, the following question arose: does this band correspond to a latent oil palm viroid or to an oil palm nucleic acid whose physico-chemical characteristics are similar to those of viroids ? Further studies of the viroid-type band of oil palms in Brazil were conducted by Beuther et al. [46]. The electrophoretic profile of this band in a temperature gradient is nearer to that of dsRNA than to that of viroids [47, 48]. Furthermore, dsRNA-specific monoclonal antibodies give a positive immunoblotting reaction [46]. Similar results suggesting the possible nature of dsRNA were obtained recently using the viroid-type band of oil palms in Ecuador [49].

Analysis of double-stranded RNA for virus diagnosis following the method developed by Morris and Dodds [50] was conducted on diseased palms from Ecuador. Numerous bands (a dozen or so), with a wide range of molecular weights and the dsRNA type of which was confirmed by their varying sensitivity to low or high ionic strength RNase, were identified in both diseased and healthy oil palms [51]. No relationship was established between any of these bands and the 'viroid-type band'.

It is seen, therefore, that while research on using nucleic acids for pathogen identification bore fruit in at least two cases—Cadang-cadang and CFD—it can sometimes involve long and difficult work using delicate molecular biology techniques. The further one goes into new fields on a molecular level, the greater the care one must take while interpreting results.

VECTOR IDENTIFICATION BY INSECT INTRODUCTIONS *EN MASSE* IN CAGES

As we have seen, direct identification of the pathogen of palm diseases of unknown origin encounters a certain number of obstacles. Among these, the impossibility of mechanically transmitting the disease from one plant to another, so as to have material available permanently to study or check pathogen existence in purified extracts, is a major handicap.

Vector identification can be a major boost for etiological studies, epidemiology, and developing control methods. However, vector detection is not merely a formality. Indeed, at the start of the 1970s, no palm disease vector insects had been identified. The following examples analyze the factors that have increased knowledge of how diseases of unknown origin are transmitted and draw out information useful for this type of study.

Blast and Dry Bud Rot

Blast, the main oil palm nursery disease in West Africa, which also affects coconut, was first identified in 1939 and until the early 1970s was attributed to a mixed fungal infection. Faced with the near-impossibility of reproducing the disease with fungi and with the knowledge that only nursery shading reduced blast percentages, a new research program was undertaken in 1973 to investigate the effects of shading, and to decide whether it was positive in itself—as shade—or as a physical barrier, against insects, for example [3, 52]. The first simple experiment consisted in comparing blast rates on (1) unprotected trees in the nursery, (2) trees entirely protected from insects in insect-proof cages, (3) trees enclosed in a cage in which insects gathered from weeds around the nurseries and from nursery palms not involved in experiments were introduced. Results show (1) the absence of blast on trees protected against insects and (2) high blast percentages on trees under normal nursery conditions or on trees stung by insects in cages, revealing the role of insects in blast appearance for the first time [52].

After several cage trials involving various species in the same cage, *Jassides* species were identified as suspected vector species. Separation of the various *Jassides* suspects (one species per cage) then revealed the vector role of *Recilia mica* Kramer (*Deltocephalinae*) [53].

This discovery rests on a simple but progressive protocol:
(1) Introduction into a single cage, each day, of all the stinging insects gathered in the nursery or on plants surrounding it, until disease symptoms appear in the cage. This is based on 'forced natural

transmission', avoiding the need for theoretical calculations and uncertain manipulations as regards forced acquisition access on diseased plants, incubation and inoculation periods, which can considerably reduce the number of insects and their lifespans.

(2)   Once insect transmission is proved, increasingly selective introductions—under the same conditions—by family and then by species.

Another factor for success may lie in the technique used to capture the insects, since butterfly nets often physically damage them. The insects were therefore gathered individually, by hand, using a tube. This technique is more labor-intensive for mass introductions (100 to 200 *Recilia* individuals introduced per day) but the insects are in much better condition.

The possibility of inducing blast on young trees in cages in this way enabled progress to be made in the etiological study as well as in vector identification.

A study by electron microscopy proved difficult, since the rot develops rapidly, and only numerous bacteria were seen [54].

However, Madagascar periwinkles subjected to *Recilia* stings which transmitted blast to oil palm show symptoms of yellowing with which mycoplasmas are associated [3, 54]. In addition, using hydroponic cultivation in cages, a comparative antibiotics trial comprising three treatments was done:

(1)   oil palm + nutritive solution + *Recilia* introductions (control)

(2)   oil palm + nutritive solution with tetracycline + *Recilia*

(3)   oil palm + nutritive solution with penicillin + *Recilia*

The results confirmed the etiological hypothesis of mycoplasmas.

In fact, tetracycline prevented blast symptoms from appearing, while penicillin only reduced the number of cases of blast compared with the control (probably due to effect of penicillin on secondary bacterial-type contaminants, leading to possible disease remission, as sometimes seen in the nursery). Resuming the ultrastructural study of blast on oil palm at as early a stage as possible in fact revealed mycoplasma-like organisms in diseased trees [3].

The same strategy also enabled identification of the vectors of dry bud rot affecting young coconuts in the Ivory Coast: *Sogatella kolophon* Kirkaldy and *S. cubana* Crowford (*Delphacidae*) [55, 56]. However, while it was seen that *Sogatella*, dry bud rot vectors in the Ivory Coast, can transmit reoviruses to crabgrass, no virus particle was identified under the electron microscope in coconut palms affected by dry bud rot [57]. However, the etiological hypothesis of a virus remains the most likely alternative, since tetracycline has no effect on the disease [3], and the

Sogatella genus includes several vector species, carrying rhabdo- and reoviruses.

Coconut Foliar Decay

Coconut foliar decay was described earlier. At the same time as the etiological research, work was carried out on the vector in Vanuatu, on two fronts:
(1) inventory of suspect species with comparison of entomofauna in unaffected and focus areas,
(2) transmission trials in cages using mass insect introductions—multispecific and monospecific—until the disease is obtained, using trees in an insect-proof cage, without insects as control.
Each cage contained 16 plants that had spent four months in the nursery. Between 60,000 and 90,000 insects were introduced into the various cages over a period of eight months. Eight months after the first introductions, the trials revealed the vector role of *Myndus taffini* Bonfils (*Homoptera Cixiidae*) [58]. Three major factors probably contributed to the success of this trial:
(1) The fact that the study comparing healthy areas and diseased areas rapidly led to *Myndus taffini* being suspected. This was largely proved by a study of a new, rapidly developing focus, where decreasing the number of *Myndus taffini* decreased the number of cases of CFD proportionately.
(2) The possibility of individual captures in tubes by virtue of abundant labor available.
(3) Mass insect introductions until the disease appeared.
In fact, trials carried out later showed that the percentage of naturally infectious insects was less than 5% [3, 59], and probably even less than 1% [60].

This last factor—number of insects over a long period—is probably responsible for the detection of the lethal yellowing vector, another *Cixiidae, Myndus crudus* Van Duzee [61], an insect which had been suspect for about a decade, but whose vector role had not yet been demonstrated. The first cases of lethal yellowing were observed 18 months after the first introductions, made at a rate of 850 *M. crudus* per month.

Detecting the CFD vector enabled significant etiological progress by:
(1) providing an extra argument excluding a mycoplasma origin, by antibiotic treatments on hydroponic crops in cages (tetracycline had no preventive or curative effect) [59], and
(2) producing as many diseased young coconuts as necessary for virological studies.

The discovery of the vector also had important consequences for developing control methods. The host plant, on which *M. taffini* lays its eggs and on which its larvae develop, is a wild *Hibiscus—H. tiliaceus* [58]. Eliminating *Hibiscus* around plantations could reduce inoculum pressure. Furthermore, it is now possible to test the tolerance of a given variety in a relatively short time, with reliable results, by submitting it to *Myndus* stings in cages, always placing the variety highly susceptible to CFD, Malayan red dwarf, in the same cage as a positive control [59–62].

CONCLUSION

The increased knowledge gained over the past 20 years with the appearance of new phytopathological concepts such as mycoplasmas or viroids, and improved techniques (electron microscopy, electrophoresis, and other molecular biology techniques) has made it possible to determine the origin of several oil palm and coconut diseases of unknown origin.

Nevertheless, however sophisticated the techniques and laboratory equipment might be, they are not sufficient to identify the cause of a disease. It is obvious, for example, that electron microscopy now plays the same role as light microscopy in the early days of phytopathology. However, as Bachelard said (in *La formation de l'esprit scientifique*, 3rd edition, Paris, J. Vrin, 1957, p. 242), 'it should not be forgotten that the microscope is an extension of the mind, and not of the eye'. Hence, it is highly likely that several researchers saw mycoplasmas under the microscope before 1967 but failed to notice them and hence to link them to the disease they were studying, since mycoplasmas were not on the list of established concepts. On the other hand, there is no point in having very good equipment and the best techniques if the study material does not initially contain the pathogen. It is therefore absolutely essential to have extensive knowledge of the pathological syndrome—and if possible of the plant—so as to take the best samples, at the right stage of the disease. This has been proved in several cases of diseases of unknown origin affecting coconut (lethal yellowing, Kaincopé, hartrot, etc.) or oil palm (Marchitez).

However, there are still numerous oil palm and coconut diseases of unknown origin (Table 6.1). Some have been the subject of only preliminary research. For others, research has provided certain clues, but their etiology remains a mystery, as with coconut dry bud rot and coconut root wilt in Kerala [63], and further study is essential. Lastly, despite extensive research, the origin of certain diseases, such as oil palm bud rot in South America, is still unknown. It is possible that other

Table 6.1 Coconut and oil palm diseases of unknown origin

Coconut		Oilpalm	
Disease	Country	Disease	Country
Thatipaka disease	India	Frond rot	Thailand
Thanjavur wilt	India	Wilt	Thailand
(or wilt of Tamil Nadu)		Turbo disease	Colombia
Leaf scorch	Sri Lanka	Bud rot	Ecuador
Natuna wilt	Indonesia		Colombia
Sarawak wilt	Malaysia	Fatal yellowing	Brazil
Yellow ringspot	Philippines	Tala seca	Brazil
Boang disease	Philippines	Ringspot disease	Peru
Socoro wilt	Philippines		Ecuador
Frond drop	Philippines		Colombia
Bristle top	Guam	Bud rot-Little leaf	Zaïre
Porroca	Colombia		

strategies, such as a comparative study of dsRNA, which was hardly even looked at in the study of bud rot, or a comparative study of proteins in diseased and healthy plants to detect viral capsid proteins, may yet lead to increased knowledge of these diseases of unknown origin.

REFERENCES

1. Doi, Y., M. Teranaka, K. Yora and H. Asuyama. Mycoplasma- or PLT group-like microorganisms found in the phloem elements of plants infected with mulberry dwarf, potato witches 'broom, aster yellows, or Paulownia witches' broom. *Ann. Phytopathol. Soc. Jpn.*, 33, 259, 1967.
2. Diener, T.O. Potato spindle tuber 'virus'. IV. A replicating, low molecular weight RNA. *Virology*, 45, 411, 1971.
3. Dollet, M. Recherches étiologiques sur les syndromes pathologiques des oléagineux tropicaux pérennes (cocotier et palmier à huile). Thèse d'Etat, Université des Sciences et Techniques du Languedoc, Montpellier, 1985.
4. McCoy, R.E., F.W. Howard, J.H. Tsai, H.M. Donselman, D.L. Thomas, H.G. Basham, R.F. Atilano, F.M. Eskafi, L. Britt and M.E. Collins. *Lethal Yellowing of Palms*, Institute of Food and Agricultural Sciences, University of Florida, Bull. 834, 1983.
5. Santana-Quadros, A. A 'doença desconhecida' do coqueiro na Zambézia. *Rev. Agric.*, 148, 33, 1972. Revista Agricola Mocambique 14, 33-34.
6. Beakbane, A.B., C.H.W. Slater and A.F. Posnette. Mycoplasmas in the phloem of coconut, *Cocos nucifera* L., with Lethal Yellowing disease. *J. Hortic. Sci.*, 47, 265, 1972.
7. Plavsic-Banjac, B., P. Hunt and K. Maramorosch. Mycoplasma-like bodies associated with Lethal Yellowing disease of coconut palms. *Phytopathology*, 62, 298, 1972.
8. Dabek, A.J.C., G. Johnson and H.C. Harries. Mycoplasma-like organisms associated with Kaincopé and Cape St. Paul wilt disease of coconut palm in West Africa. *PANS*, 22, 354, 1976.
9. Dollet, M., and J. Giannotti. Maladie de Kaïncopé: présence de mycoplasmes dans le phloème de cocotiers malades. *Oléagineux*, 31, 169, 1976.

10. Dollet, M., J. Giannotti, J.L. Renard and S.K. Ghosh. Etude d'un jaunissement létal du cocotier au Cameroun: la maladie de Kribi. Observations d'organismes de type mycoplasmes. *Oléagineux*, 32, 317, 1977.

11. Nienhaus, F., M. Schuiling, G. Gliem, U. Schinzer and A. Spittel. Investigations on the etiology of the lethal disease of coconut palm in Tanzania. *PflKrankh*, 89, 185, 1982.

12. Steiner, K.G., F. Nienhaus and K.J. Marschall. Rickettsia-like organisms associated with a decline of coconut palm in Tanzania. *PflKrankh*, 84, 345, 1977.

13. McCoy, R.E. Effect of various antibiotics on development of lethal yellowing in coconut palm. *Proc. Florida State Hortic. Soc.*, 86, 503, 1973.

14. McCoy, R.E. Remission of lethal yellowing in coconut palm treated with tetracycline antibiotics. *Plant Dis. Rep.*, 56, 1019, 1972.

15. Steiner, K.G. Remission of symptoms following tetracycline treatment of coconut palms affected with Kaincopé disease, *Plant Dis. Rep.*, 60, 617, 1976.

16. Dollet, M. Maladies d'origine inconnue du palmier à huile au Pérou et en Equateur, IRHO Report, 1976.

17. Parthasarathy, M.V., W.G. van Slobbe and C. Soudant. Trypanosomatid flagellate in the phloem of diseased coconut palms. *Sciences*, 192, 1346, 1976.

18. Dollet, M., J. Giannotti and M. Ollagnier. Observation de protozoaires flagellés dans les tubes criblés de palmiers à huile malades, *C.R. Acad. Sci. Paris*, Sér. D, 284, 643, 1977.

19. Waters, H. A wilt disease of coconuts from Trinidad associated with *Phytomonas* sp. a sieve tube restricted protozoan flagellate. *Ann. Appl. Biol.*, 90, 293, 1978.

20. Dollet, M., and G. Lopez. Etude sur l'association de protozoaires flagellés à la Marchitez sorpresiva du palmier à huile en Amérique du Sud. *Oléagineux*, 33, 209, 1978.

21. Menara, A., M. Dollet, D. Gargani and C. Louise. Culture in vitro sur cellules d'invertébrés des *Phytomonas* sp. (Trypanosomatidae) associés au Hartrot, maladie du cocotier, *C.R. Acad. Sci. Paris*, Sér. III, 307, 597, 1988.

22. Louise, C., M. Dollet and D. Mariau. Recherches sur le Hartrot du cocotier, maladie à *Phytomonas* (Trypanosomatidae) et sur son vecteur *Lincus* sp. (Pentatomidae) en Guyane. *Oléagineux*, 41, 437, 1986.

23. Hatta, T., and R.I.B. Francki. Enzyme cytochemical method for identification of cucumber mosaic virus particles in infected cells. *Virology*, 93, 265, 1979.

24. Hatta, T., and R.I.B. Francki. Identification of small polyhedral virus particles in thin section of plant cells by an enzyme cytochemical technique. *J. Ultrastruct. Res.*, 74, 116, 1981.

25. Bigornia, A.E. Evaluation and trends of researches on the coconut Cadang-cadang diseases. *Philipp. J. Coconut Studies*, 2, 5, 1977.

26. Zelazny, B., J.W. Randles, G. Boccardo and J.S. Imperial. The viroid nature of the Cadang-Cadang disease of coconut palm. *Scienta Filipinas*, 2, 46, 1982.

27. Randles, J.W. Detection in coconut of rod-shaped particles which are not associated with disease, *Plant Dis. Rep.*, 59, 349, 1975.

28. Randles, J.W. Association of two ribonucleic acid species with Cadang-Cadang disease of coconut palm. *Phytopathology*, 65, 163, 1975.

29. Haseloff, J., N. Mohamed and R.H. Symons. Viroid RNAs of the Cadang-Cadang disease of coconuts. *Nature* (London), 29, 317, 1982.

30. Randles, J.W., G. Boccardo, M.L. Retuerma and E.P. Rillo. Transmission of the RNA species associated with Cadang-Cadang of coconut palm, and the insensitivity of the disease to antibiotics. *Phytopathology*, 67, 1211, 1977.

31. Mohamed, N.A., R. Bautista, G. Buenaflor and J.S. Imperial. Purification and infectivity of the coconut Cadang-Cadang viroid. *Phytopathology*, 75, 79, 1985.

32. Boccardo, G., R.G. Beaver, J.W. Randles and J.S. Imperial. Tinangaja and bristle top, coconut diseases of uncertain etiology in Guam, and their relationship to Cadang-Cadang disease of coconut from the Philippines. *Phytopathology*, 71, 1104, 1981.

33. Keese, P., M.E. Osorio-Keese and R.H. Symons. Coconut Tinangaja viroid: sequence homology with coconut Cadang-Cadang viroid and other potato spindle tuber viroid related RNAs. *Virology*, 162, 508, 1988.

34. Randles, J.W., G. Boccardo and J.S. Imperial. Detection of the Cadang-Cadang associated RNA in African oil palm and buri palm. *Phytopathology*, 70, 185, 1980.

35. Dollet, M., D. Gargani and G. Boccardo. Recherches sur l'étiologie d'un dépérissement des cocotiers au Vanuatu, examen en microscopie électronique et comparaison des acides nucléiques d'arbres sains et arbres malades par électrophorèse sur gel de polyacrylamide. Paper presented in Int. Conf. Trop. Crop. Prot., Lyons, p. 65, Programme Abstracts, July 8–10, 1981.

36. Julia, J.F., M. Dollet, J. Randles and C. Calvez. Le dépérissement foliaire par *Myndus taffini* (DFMT): nouveaux résultats. *Oléagineux*, 40, 19, 1985.

37. Randles, J.W., J.F. Julia, C. Calvez and M. Dollet. Association of single-stranded DNA with the foliar decay disease of coconut palm in Vanuatu. *Phytopathology*, 76, 889, 1986.

38. Randles, J.W., D. Hanold and J.F. Julia. Small circular single-stranded DNA asociated with foliar decay disease of coconut palm in Vanuatu, *J. Gen. Virol.*, 68, 273, 1987.

39. Julia, J.F. *Myndus taffini* (*Homoptera cixiidae*), vecteur du dépérissement foliaire des cocotiers au Vanuatu. *Oléagineux*, 37, 409, 1982.

40. Randles, J.W. and D. Hanold. Coconut foliar decay virus particles are 20 nm icosahedra. *Intervirology*, 30, 177, 1989.

41. Turner, P.D. *Oil Palm Diseases and Disorders*, Kuala Lumpur: Oxford University Press, p. 167, 1981.

42. Dollet, M., C. Saussol and D. Gargani. La pourriture du coeur du palmier à huile (*Elaeis guineensis*) dans la région de l'Oriente en Equateur, IRHO report, 1984.

43. Schumacher, J., J.W. Randles and D. Riesner. A two dimensional electrophoretic technique for the detection of circular viroids and virusoids. *Anal. Biochem.*, 135, 288, 1983.

44. Singh, R.P., A.C. De Avila, A.N. Dusi, A. Boucher, D.R. Trindade, W.G. van Slobbe, S.G. Ribeiro and M.E.N. Fonseca. Association of viroid-like nucleic acids with the fatal yellowing disease of oil palm. *Fitopatol. bras.*, 13, 392, 1988.

45. Mazzolini, L., D. Dambier and M. Dollet. Mise en évidence d'une molécule de type viroïde dans le palmier à huile en Equateur. *Proc. Secondes rencontres de virologie végétale CNRS-INRA*, Aussois, 1989.

46. Beuther, E., W.G. van Slobbe and D. Riesner. Detection of double-stranded RNA in oil palms as an indication for a virus infection. *Proc. IVth Int. plant virus epidemiol. work*, p. 342. Montpellier, 1989.

47. Rosenbaum, V., and D. Riesner. Temperature-gradient gel electrophoresis. Thermodynamic analysis of nucleic acids and proteins in purified form and in cellular extracts. *Biophys. chem.*, 26, 235, 1987.

48. Riesner, D. Physical-chemical properties. Structure formation. In: *The Viroids*. T.O. Diener, ed. New York and London: Plenum Press, p. 63, 1987.

49. Bernard, V., and E. Beuther, unpublished data, 1990.

50. Morris, T.J., and J.A. Dodds. Isolation and analysis of double-stranded RNA from virus-infected plant and fungal tissue. *Phytopathology*, 69, 854, 1979.

51. Mazzolini, L., and M. Dollet, unpublished data, 1988.

52. Renard, J.L., D. Mariau and P. Quencez. Le Blast du palmier à huile: rôle des insectes dans la maladie. Résultats préliminaires. *Oléagineux*, 30, 497, 1975.

53. Desmier de Chenon, R. Mise en évidence du rôle de *Recilia mica* Kramer (Homoptera, Cicadellidae, Deltocephalinae) dans la maladie du Blast de pépinières de palmiers à huile en Côte d'Ivoire. *Oléagineux*, 34, 107, 1989.

54. Dollet, M. Research on the etiology of Blast of oil and coconut palms. *Proc. 4th meeting Int. Counc. Lethal Yellowing*, Univ. of Florida, Agr. Res. Cent., Ft. Lauderdale, Publ. FL-80-1, p. 19, 1980.

55. Julia, J.F. Mise en évidence et identification des insectes responsables des maladies juvéniles du cocotier et du palmier à huile en Côte d'Ivoire. *Oléagineux*, 34, 385, 1979.

56. Julia, J.F. and D. Mariau. Deux expèces de *Sogatella* (Homoptère Delphacidae) vectrices de la maladie de la Pourriture sèche du coeur des jeunes cocotiers en Côte d'Ivoire. *Oléagineux*, 37, 517, 1982.

57. Sumardiyono, Y. Contribution à l'étude des maladies du cocotier à hypothèse étiologique virale, exemple de la Pourriture sèche du coeur. IIIe cycle thesis, Université des Sciences et Techniques du Languedoc, Montpellier, 1984.

58. Julia, J.F. *Myndus taffini* (*Homoptera Cixiidae*) vecteur du dépérissement foliaire des cocotiers au Vanuatu. *Oléagineux*, 37, 409, 1982.

59. Julia, J.F. M. Dollet, J.W. Randles and C. Clavez. Le dépérissement foliaire du cocotier par *Myndus taffini* (DFMT): nouveaux résultats. *Oléagineux*, 40, 19, 1985.

60. Julia, J.F., and M. Dollet, unpublished data, 1984.

61. Howard, F.W., R.C. Noris and D.L. Thomas. Evidence of transmission of palm lethal yellowing agent by a planthopper, *Myndus crudus* (Homoptera: Cixiidae). *Trop. Agric. (Trinidad)*., 60, 168, 1983.

62. Calvez, C., J.F. Julia and M. de Nucé de Lamothe. L'amélioration du cocotier au Vanuatu et son intérêt pour la région du Pacifique. Rôle de la station de Saraoutou. *Oléagineux*, 40, 477, 1985.

63. Solomon, J.J., M.P. Govindankutty and F. Nienhaus. Association of mycoplasmalike organisms with the coconut root (wilt) disease in India. *PflKrankh*, 90, 295, 1983.

Chapter 7

A QUARTER CENTURY EXPERIENCE ON COCONUT DISEASES OF UNKNOWN ETIOLOGY

Luigi Chiarappa

Luigi Chiarappa, after retiring from Food and Agriculture Organization of the United Nations, is a Consultant in Plant Pathology and International Agriculture. He holds a 'Laurea' degree in Agricultural Sciences from the University of Florence, Italy and a PhD in Plant Pathology from the University of California, Davis. He served as: Agronomist, ICLE, Chile 1950–1951; Assistant Entomologist, Di Giorgio Fruit Corp. 1952–1956; Research Plant Pathologist, Di Giorgio Corp. 1958–1962; Tropical Plant Pathologist (1962–1976), Senior Plant Pathologist (1976–1982) and Chief, Plant Protection (1982–1986), FAO, Rome, Italy. American Phytopathological Society (Emer.); Mediterranean Phytopathological Union; ICVG. Research interests: diseases of tropical and subtropical crops, epidemiology, crop loss appraisal. Address: 4221 Montgomery Ave., Davis, California, 95616, USA.

CONTENTS

IF COCONUTS WERE APPLES

If coconuts were apples, with all probability there would not be a workshop of this kind in India and we would not be speaking about unknown diseases of coconuts. Instead, we would be meeting in some temperate country and, most likely, we would be speaking about apple diseases. But do apples deserve to be considered the most diseased crop in the world? This could be true if we were to accept the results of a survey of phytopathological literature which not long ago identified apples as having the largest number of published papers. This seems also to apply to recent specialized literature on virus and virus-like diseases. For example, the latest book by Friedlund lists 22 apple diseases due to virus-like pathogens as against 14 described by Roistacher for citrus and only 2 for coconuts [1, 2]. This, indeed, appears to be a very unusual situation. The purpose of this paper is mainly to focus attention on the importance of coconuts in the tropics and on the seriousness of some diseases of uncertain etiology. My long association with FAO suggested the title of this contribution.

During nearly a quarter of century with that organization I was able to focus priority attention on two major areas: the status of plant pathology in developing countries and the socioeconomic importance of local plant diseases. Within these, I was particularly motivated by diseases of unknown etiology and international significance. All this started with a worldwide survey of coconut diseases, which had been described in the literature either incompletely or incorrectly. There was need for a comparative study of these diseases to be carried out with the same methodology and timing. Dr. Karl Maramorosch was selected for this assignment for three main reasons: he knew coconuts, having served on the FAO Cadang-cadang Project in the Philippines, he was an experienced world traveler, and he had the reputation of being an excellent photographer. The survey was organized and conducted in record time. The results were useful to clarify obscure nomenclatures, to identify symptomatological similarities, and to actually redirect some etiological research [3].

COCONUT PRODUCTION AND UTILIZATION

Total world area in coconut production is not known with certainty. The FAO estimates that there are over 6 million hectares of plantings of which 90% are in Asia and Oceania. In the tropics coconuts (*Cocos nucifera*) are considered the tree of heaven because of their many uses. At least 50% of the total production is consumed fresh and contributes to the diet of millions of people. Coconut oil is very important in the manufacture of soap, margarine, and glycerol (much of which is used for explosives). Desiccated nuts find use in candies and cakes, and also in cattle and poultry feeds. Coconut byproducts have almost no limitations in use. Coir fiber and coir products derived from the husks form the basis of many small industries in the tropics. For example, in Kerala in India, over half a million people depend on the coir industry. Millions of tons of shells are used for making bowls, latex collecting cups, scoops, and even smoking pipes.

Shells are also used as fuel either directly or as charcoal. This, in turn, is good for the production of activated carbon for gas absorption, bleaching, or similar chemical uses. Coconut trunks are employed in building houses and sheds. Leaves are used for thatching and for making screens and baskets. Leaflet midribs find use in the making of fish and lobster traps or stiff brooms. Roots are used for medicinal concoctions against fever or dysentery, or for mouthwashes and gargles. Finally, tapped coconut inflorescences produce toddy, a sugary juice that can be fermented into arrak or made into vinegar. In addition it must be emphasized that coconuts in the tropics are grown where no other crops can be raised, and that coconuts are the typical small grower's crop requiring little or no capital inputs. For this reason considerable attention has been given in recent years to the improvement of coconut breeding and agronomy. Unfortunately, a serious obstacle to progress in both areas has been the existence of some very important diseases of unknown etiology (Plates 7.1, 7.2).

DISEASES

Three diseases will be reviewed in this paper: lethal yellowing, Cadang-cadang, and the complex variously named Malayan, Sarawak or Natuna wilt.

Lethal Yellowing

According to recent reviews, over 80% of the 4.3 million palms growing in Jamaica and more than 50% of 1 million palms in Florida have succumbed to lethal yellowing [4, 5]. No figures are available for the West Coast of Africa, where the same disease is known as Cape St. Paul wilt or Kaincopé [6]. The disease is present in Togo, Cameroon, Dahomey, and Nigeria. Recently, the disease in Ghana has been reported to be only 50 km away from the border with the Ivory Coast, where a large international coconut improvement program is under way [7]. A disease similar to lethal yellowing, but due to a different pathogen, also exists in East Africa and attempts are being made to coordinate work on this problem with research in Ghana and Mexico [7]. In Mexico the disease has reached epidemic proportions in the Yucatan peninsula with more than half a million palms affected, mainly in the touristic coastal areas. The Mexican copra-producing districts are still free from lethal yellowing but they could soon become affected because of the spread capabilities of the insect vector and the prevailing susceptibility of palms grown in these districts. A national plan against the disease in Mexico has been prepared and recently revised [8].

Lethal yellowing is so called because of its bright or butter-yellow discoloration of the leaves. This symptom is typical of the highly susceptible Jamaica or West Africa tall coconut. In the Panama talls, Malayan dwarfs, and, possibly, in other more resistant palms the disease does not produce yellowing but only a progressive discoloration and bronzing of the leaflets (Plates 7.3, 7.4) [9]. This symptom is accompanied by a gradual wilting and drying of the lower whorls, which remain vertically suspended from the apex of the trunk. Frequently the midribs of some of the younger fronds break halfway (Plate 7.5). Other typical symptoms in susceptible coconuts are necrosis of inflorescences (Plate 7.6) and premature dropping of the nuts in all stages of development. Symptoms in some palms (such as *Corypha alata, Pritchardia* spp.) are similar to those in Jamaica tall coconuts. In other palms (such as *Veitchia merrillii, Phoenix* spp.) the leaves turn brown with water-soaked marks along the pinnae. Breaking of the midribs and of the leaf base also occurs.

Lethal yellowing is known to affect at least 30 palm species and to have possibly one non-palm host, *Pandanus utilis* [10]. Within *Cocos nucifera* considerable differences in susceptibility exist. This is possibly due to the different origin and evolution of two recognizable taxonomic groups [11].

Mycoplasma-like organisms (MLOs) have been observed by several investigators in diseased palms from Jamaica, Florida, Togo, and Cameroon [12, 13, 14]. Further evidence supporting MLOs as etiological agents has been derived from tetracycline injection experiments [15, 16, 17]. In Tanzania, rickettsia-like organisms have been observed [18].

The first successful insect transmission of lethal yellowing was observed by Heinze and Schuiling in Jamaica [19]. These workers obtained symptoms on caged palms on which mixtures of *Myndus crudus* and other insects collected from the wild were placed. Because of the very low rate of transmission and the presence of other insects, these results were discounted. Subsequent studies in Florida showed instead that wild populations of *M. crudus* were able to transmit lethal yellowing to *Veitchia merrillii*. Subsequent attempts to transmit the disease with this same insect reared from laboratory colonies and from wild populations were unsuccessful [20].

So far, the only practical control is through the use of resistant palms. These include Panama talls, Malayan dwarfs, and their hybrids ('Maypan'). Extensive planting of these coconuts have been made in both Jamaica and Florida. Recently, high losses have been reported from Florida where the problem appears to be confined to hotels and golf courses and to a number of palm off-types.

Cadang-cadang

Under many considerations, Cadang-cadang can be considered of less economic importance than lethal yellowing. The latter is a killing disease of palms of nearly all ages causing rapid loss of production. Cadang-cadang is a debilitating disease causing slow decline, having no effect on palms less than 10 years old. In mature palms (22 years and older) the disease takes an average of 7.5 years to complete its course. In older palms (44 years and older) it will take twice as much time. Lethal yellowing is also more important because of its wider geographic distribution and greater quarantine risk. It occurs in Africa, the Caribbean, Mexico, Texas, and Florida. Cadang-cadang is limited to the central area of the Philippines and Guam. Furthermore, lethal yellowing is transmitted by at least one insect vector capable of flying over long distances while the insect vector of Cadang-cadang is not known and, possibly, does not exist. In the early 1950s incidence of Cadang-cadang ranged from 9 to 67% in different provinces of the Philippines and an estimated 30 million palms were lost. Now the disease appears to have entered an endemic stage with a spread of less than 500 meters per year and, occasionally, with no spread at all [21].

Infected bearing palms show a progression of symptoms culminating in death. In the early stage the first symptom is the rounding of the nuts and the scarification of their equatorial zone (Plate 7.7). Chlorotic leaf spots then appear on young expanded fronds while inflorescences become at first stunted then sterile. During the middle stage nut production is greatly reduced and finally ceases altogether. The late stage is characterized by the coalescence of leaf spots and foliage chlorosis, by decline in size and number of fronds, and, finally, by death.

Symptomatology, in general, is unsatisfactory for disease diagnosis because of the long delay (two to four years) between time of inoculation and symptom appearance.

Only members of the *Palmae* can be infected by artificial inoculation. These include the betel nut (*Areca catechu*), the buri palm (*Orypha elata*), the Manila palm (*Adonidia merrillii*), the African oil palm (*Elaies guineensis*), the palmera (*Chrysalidocarpus lutescens*), and the date palm (*Phoenix dactylifera*). A naturally infected African oil palm has been also detected. The viroid nature of Cadang-cadang has been established [22]. Four circular forms of viroid can be extracted from diseased palms (CC1S, CC1L, CC2S, CC2L). There is evidence for some unique modifications of the viroid primary structure with time and disease progress [23]. The search for a vector of Cadang-cadang has been carried out unsuccessfully for nearly three decades. The most recent list of insects associated with infected and healthy palms in southeastern Luzon is that by Zelazny and Pacumbaba [24].

The possibility that humans could be the vector through the use of infected knives has been recently suggested [25].

Sanitation (e.g. elimination of sources of inoculum and possible disinfection of knives or other cutting tools) could effectively reduce disease spread. Replanting with seed from disease-free areas also appears effective especially if done in conjunction with a shortened (10 to 15 years) production cycle.

Malayan, Sarawak and Natuna Wilts

The economic importance of Malayan, Sarawak, and Natuna wilts varies greatly from location to location and with the size of the outbreak areas. Geographic distribution appears to be limited to the Malayan peninsula, Borneo, Indonesia, the Philippines, and, possibly, some Pacific islands. Since the etiology of all these wilts is still obscure, they will be treated for the time being as a single group based on common epidemiological and symptomatological features.

All wilts appear to develop to a greater or lesser extent in a protean fashion: bulging here and shrinking there like an amoeba [26]. This means that in any given tropical location, due to some changes in the locally endemic pathosystem, there is a sudden disease flare-up with consequent death and disappearance of susceptible individuals. This is then followed by a period or relative recession when the disease is reduced below detection levels and is no longer seen, as in the Keta area of Sarawak [27]. An endemic pathosystem implies the coexistence and continuous association of the pathogen and its host in a situation in which parasitism never stops [28]. For obligate parasites, such as viruses and other procariotic organisms, the amount of disease may constantly vary, but it never reaches the zero level leading to extinction of the parasite.

As already indicated, this group of diseases appears to share the same epidemiological characteristics. As for symptomatology, these wilts are all very similar in producing symptoms like those described for the Panama tall coconut affected by lethal yellowing (Plates 7.4, 7.5, 7.6).

CONCLUSIONS

In looking back at nearly a quarter of century of work on coconut diseases of unknown etiology it can be said that research advances have been made only on Cadang-cadang and lethal yellowing. However, much more fundamental knowledge is required for both diseases to better understand obscure aspects of their pathogenesis. Much more information is particularly needed for lethal yellowing especially in vector biology and epidemiology as well as in genetic resistance. Without this knowledge it is practically impossible to develop durable forms of control and reduce crop vulnerability in the tropics. In other diseases of unknown etiology, practically no progress has been made in the last 25 years. As already stressed, coconuts are not apples and as such they do not attract much research effort. It is difficult to see that this situation will change in the future. Under these circumstances, the best to be hoped is to exploit the experience of the past to achieve better results in the future. Based on this experience there appear to be at least three needs to satisfy and three approaches to follow.

The Need to Keep Open Minds

Most of the FAO research efforts in the Philippines and Jamaica were directed only at viruses and at potential vectors. This one-track period lasted about 20 years in the Philippines and 6 years in Jamaica. It was not until after science showed that MLOs and viroids could cause disease in higher plants that research targets in both countries were changed and breakthroughs obtained.

   This means that for the other coconut diseases still of unknown etiology we should be prepared to look for pathogenic agents yet unknown to science.

The Need for Outside Assistance

Research under primitive conditions in developing countries can be extremely difficult and, most often, unfruitful. This is particularly true when sophisticated techniques and equipment are required. For example, it took nearly three years in Jamaica to properly operate a newly acquired electron microscope and at least 10 years in the Philippines before suitable research facilities were operational. Under these conditions it is essential to secure valid foreign assistance and ensure its continuity to the long-term research effort.

The Need for International Collaboration

Sequencing of FAO experts on short coconut research assignments proved ineffective in both Jamaica and the Philippines because of scientific isolation, segmentation of efforts, and lack of a critical mass. The clue to success was the establishment of solid and continuous linkages between the local research and that conducted in advanced laboratories in Australia, the United States, and Europe. This was achieved in the Philippines by establishing a long-term collaboration agreement with the Waite Agricultural Research Institute and in Jamaica by the creation of ICLY the International Council on Lethal Yellowing.

In light of this experience, we should now consider the future. Many important coconut diseases of unknown etiology still remain. In addition to Malayan, Sarawak, and Natuna wilts, there are a number of serious diseases whose etiology remains undetermined. These include Kerala wilt of India and 'la maladie' of the New Hebrides. It is important that greater resources be directed at these problems. As for other diseases, and in particular lethal yellowing, many areas still require attention. For example, speed and reliability of detection methods must be improved. The inherent difficulty of working with tall palms only in open spaces (Plate 7.8) must be overcome by finding alternate host plants suitable for studies under more controlled conditions. The genetics and cytogenetics of the coconut palm also necessitate considerable study. Parallel to this, improved methods for breeding, for evaluating genetic resistance, and for rapid tissue culture propagation are all essential.

Clearly, in all these areas there is ample room for basic and applied research involving many disciplines including entomology, plant pathology, genetics, plant breeding, physiology, and biochemistry. Procuring and cementing together these disciplines in a viable research program is not an easy task in any one country. However, it would seem possible to approach this objective by closely linking national undertakings to an international scheme able to attract the interest, consciousness, and enthusiasm of the world scientific community.

Mexico is now launching a long-term national program on lethal yellowing. To assist Mexico in this serious endeavour, the re-activation of the International Council on Lethal Yellowing could mobilize again the creative powers of the international scientific community and secure for the future new and still unthinkable results.

REFERENCES

1.   Friedlund, P.R. *Virus and virus-like diseases of pome fruits and simulating noninfectious disorders.* Coop. Ext. Wash. State Univ., Pullman SP0003, 1989.

2. Roistacher, C.N. *Handbook for Detection and Diagnosis of Graft-transmissible Diseases of Citrus.* Rome. FAO (In press.)

3. Maramorosch, K. A. *Survey of Coconut Diseases of Unknown Etiology.* Rome: FAO 1964.

4. Tsai, J.H., and K. Maramorosch. Lethal yellowing of palms. In: *Mycoplasma Diseases of Woody Plants.* Raychaudhuri and Rishi, eds. New Delhi: Malhotra Publ. House, p. 29, 1988.

5. McCoy, R.E., F.W. Howard, J.H. Tsai, H.M. Donselman, D.L. Thomas, H.G. Basham, R.A. Atilano, F.M. Eskafi, L. Britt and M.E. Collins. *Lethal Yellowing of Palms,* Univ. of Florida Tech. Bull. 834, F. Lauderdale, 1983.

6. Chiarappa, L. Co-identity of lethal yellowing with Kaincopé disease of West Africa. *Principes,* 17, 152, 1973.

7. Harries, H.C., personal communication, 1989.

8. Chiarappa, L. Consultoria sobre Amarillamento Letal del Cocotero, IICA Report, Mexico (mimeo), 1990.

9. Whitehead, R.A. *Collection of Coconut Germplasm from the India / Malaysian Region, Peru and the Seychelles Islands and Testing for Resistance to Lethal Yellowing Disease in Jamaica.* Rome: FAO, 1960.

10. Thomas, D.L., H.M. Donselman. Mycoplasma-like bodies and phloem degeneration associated with declining *Pandanus* in Florida. *Plant Dis. Rep.,* 63, 911, 1979.

11. Harries, H.C. Evolution, dissemination and classification of *Cocos nucifera. Bot. Rev.,* 44, 265, 1978.

12. Beakbane, A.B., C.W. Slater and A.F. Posnette. Mycoplasmas in the phloem of coconuts (*Cocos nucifera* L.) with lethal yellowing disease. *J. Hortic. Sci.,* 47, 265, 1972.

13. Heinze, K.G., H. Petzold and R. Marwitz. Beitrag zur Atiologie der Todlichen Vergilbung der Kokospalme. *Phytopathol. Z.,* 74, 230, 1972.

14. Plavsic-Banjac, B., P. Hunt and K. Maramorosch. Mycoplasma-like bodies associated with lethal yellowing disease of coconut palms. *Phytopathology,* 62, 298, 1972.

15. Hunt, P., A.J. Dabek and M. Schuiling. Remission of symptoms following tetracycline treatment of lethal yellowing infected coconut palms. *Phytopathology* 64, 307, 1974.

16. McCoy, R.E. Remission of lethal yellowing in coconut palm treated with tetracycline antibiotics. *Plant Dis. Rep.,* 56, 1019, 1972.

17. Steiner, K.G. Remission of symptoms following tetracycline treatment of coconut palms affected with Kaincopé disease. *Plant Dis. Rep.,* 60, 617, 1976.

18. Steiner K.G. Suspected lethal yellowing disease of coconut palms in Tanzania. *FAO Plant Prot. Bull.,* 16, 10, 1978.

19. Heinze, K.G., and M. Schuiling. 10th Report Res. Dept. Coconut Industry Board, Kingston, Jamaica, 1970.

20. Howard, F.W., and D.L. Thomas. Transmission of palm lethal decline to *Veitchia merrillii* by plant hopper *Myndus crudus. J. Econ. Entomol.,* 73, 715, 1980.

21. Zelazny, B. Distribution and spread of cadang-cadang disease of coconut palms in the Philippines. *Acta Phytopathol. Acad. Sci. Hung.,* 14, 115, 1979.

22. Randles, J.W. Coconut cadang-cadang viroid. In: *Subviral Pathogens of Plants and Animals: Viroids and Prions.* K. Maramorosch and J.J. McKelvey, eds., New York: Academic Press, pp. 39–74, 1985.

23. Randles, J.W. Coconut cadang-cadang. In: *The Viroids.* T.O. Diener, ed. Plenum Press, p. 265, 1987.

24. Zelazny, B., J.W. Randles, G. Boccardo and J.S. Imperial. The viroid nature of the cadang-cadang disease of coconut palm. *Sci. Filip.,* 2, 45, 1982.

25. Maramorosch, K. The cadang-cadang disease of coconut palms. *Rev. Trop. Pl. Path.,* 4, 109, 1987.

26. Van der Plank, J.E. *Principles of Plant Infection.* New York and London: Academic Press, 1975.

27. Chiarappa, L., and Kueh Tiong-Kheng. Sematan wilt of coconut: A disease no longer detectable. *FAO Plant Prot. Bull.*, 30, 102, 1982.

28. Robinson, R.A. *Host Management in Crop Pathosystems.* New York: McMillan Publ. Co., p. 17, 1987.

29. Chiarappa, L. The probable origin of lethal yellowing and its co-identity with other lethal diseases of coconut. *5th Session FAO Tech. Work. Party* (AGP/CNP/79/4) (mimeo), 1979.

Chapter 8

PLANT DISEASES ASSOCIATED WITH MITES AS VECTORS OF KNOWN VIRUSES AND UNKNOWN ETIOLOGICAL AGENTS

Chuji Hiruki

Chuji Hiruki is a University Professor of Plant Virology at University of Alberta, Canada. He received his BSc in Agricultural Science (Plant Pathology) 1954 and PhD (Plant Virology) 1963 at Kyushu University, Japan. Visiting Plant Pathologist 1963–1964 at University of California, Berkeley; Honorary Fellow (Biophysics) 1964–1966 at University of Wisconsin, Madison; Assistant Professor, Alberta 1966–1970; Associate Professor, Alberta 1970–1976; Professor, Alberta 1976–1991; University Professor, Alberta 1991 to date; A.G. McCalla Professor 1987; study leaves at Agricultural University, Wageningen 1972; INRA, Versailles 1972; Institute of Plant Virus Research, Japan 1973; University of Queensland, Brisbane 1984; University of Tokyo, University of Hokkaido 1985; Kyoto University 1988, Japan. Royal Society of Canada (Fellow), New York Academy of Science, Sigma Xi, Canadian Phytopathological Society (President 1990-91), American Phytopathological Society (Fellow), Phytopathological Society of Japan. Research interests: fungal transmission of plant viruses; plant diseases associated with mollicutes; molecular diagnosis of viroids, viruses, and mollicutes; genetics analyses of plant-virus interactions. Address: Department of Plant Science, University of Alberta, Edmonton, Alberta, T6G 2P5, Canada.

CONTENTS

INTRODUCTION

While progress in research on mite transmission of plant pathogens has been slow since the first evidence was reported in 1927 [1], there is a recent resurgence of interest in the subject due to the widespread and consistent occurrence of plant diseases associated with mites as vectors of known and unknown etiological agents. Since several excellent reviews are available on early research [2–8], this chapter will deal mainly with recent progress in the subject area.

MITES AS VECTORS

Members of four families of plant-feeding mites, the Tetranychidae, the Tenuipalpidae, the Tarsonemidae, and the Eriophyidae, have been investigated as vectors of plant viruses and/or disease agents of uncertain identity. All the confirmed vectors of plant disease agents are known to belong to the Eriophyidae, which consist of five subfamilies. Genera with species recognized to be vectors of plant disease agents occur in only three of them. These include *Cecidophyopsis* and *Colomerus* in the Cecidophyinae, *Eriophyes* and *Phytoptus* in the Eriophyinae, and *Abacarus, Aculus, Calacarus* and *Phyllocoptes* in the Phyllocoptinae [9]. There is substantial evidence that several species of the eriophyid mites transmit plant pathogens, at least some of which have been identified as viruses; other reported evidence requires confirmation or further consolidation.

Tetranychidae as Suspected Vectors

The Tetranychidae are active plant pests known as spider mites, since many species form a fine web over the leaves or even the entire plant. The mites are green, yellow, orange, or red. They are eight-legged, medium-sized mites (up to 0.8 mm long), oval or pear-shaped. Their piercing, sucking mode of feeding, free mobility, and wide host range suit them well to be vectors of plant viruses [4]. Sometimes the damage caused by feeding confusingly resembles symptoms of virus diseases.

The two-spotted spider mite *Tetranychus telarius* (= *T. urticae*) was reported as another vector, in addition to aphids, of potato virus Y (PVY) from potato and *Nicotiana glutinosa* to potato in 1963 [10]. Both optimal acquisition feeding time and inoculation feeding time were 5 min. About 50% transmission was reportedly achieved using potato seedlings as assay plants under such conditions. However, later investigations [11, 12] failed to confirm this report and thus no reference is made in recent reviews on PVY [13, 14].

Tenuipalpidae as Suspected Vectors

The Tenuipalpidae mites are flat, oval, dark red, and eight-legged, about 0.3 mm in length. Infestation with *Brevipalpus obovatus* Donn., false spider mite, has been associated with citrus leprosis [15, 16]. In Florida, the mites induced leprosis symptoms on citrus plants even when collected from *Bidens pilosa* L., a nonsusceptible plant, or from geographical areas where leprosis was unknown [17]. The leprosis symptoms also spread from the affected donor patch graft to receptor tissue, suggesting the presence of a virus-like pathogen [18]. In Brazil, *B. phoenicis* (Geijskes) induced leprosis in Lima sweet orange *Citrus sinensis* (L.) Osb., when they were transferred from diseased plants, or had fed previously on the lesions [19]. Virus particles measuring 100–150 nm × 40 nm occur in the nucleus and cytoplasm of the leprosis-affected citrus plants [19], and resemble those found in several orchid species [20–25] and those in association with the coffee ringspot disease in that they are rhabdovirus-like particles devoid of envelopes [19]. However, their relationship to rhabdoviruses is uncertain and it is not yet clearly established if leprosis lesions are due to a mite toxin or to a localized virus transmitted by the mite.

Tarsonemidae as Suspected Vectors

The Tarsonemidae are eight-legged, small (up to 0.3 mm long), oval-shaped mites. Rice tarsonemid mites have been reported to be associated with spherical, virus-like particles 35 nm in diameter. Their role as virus vectors has recently been suggested on the basis of the detection of serologically identical particles in rice plants that had been infested by the viruliferous mites [26]. At present, it is not known whether mites, either alone or when combined with the virus, reduce the rice growth and yield. Since the rice tarsonemid mites are usually parasitic on rice plants in tropical and subtropical areas, it is important to confirm the report and to establish optimal conditions for mite transmission.

Eriophyidae as Vectors and Their Biology

The Eriophyidae, a unique group of small mites, are represented by about 1,000 described species of phytophagous mites [27]. The mites range from 100 to 250 μm in length and are the smallest arthropods known to infest plants in several genera with highly developed host specificity [28].

The development of certain eriophyids may be completed in 6–14 days depending on environmental conditions. Two stages are known to occur for nymphal instars, the second ending in a resting period, at which time the genitalia develop. Males, usually smaller than females, are rarely found in some species. An alternate female form, specialized in hibernation, may be found in some species. Males deposit a spermatophore which is then picked up by females. The generation time of eriophyids in optimal conditions is less than a week. Thus, in temperate or subtropical climates, more than 30 generations are produced annually. Wind is a major factor in dispersing *Aceria (Eriophyes) tulipae* K. [29–31].

The eriophyids have piercing, sucking mouth parts. Their thin stylets are enclosed in a groove in the rostrum. The integument, alimentary system, and hemolymph are similar to those found in higher arthropods. The reproductive system is composed of bilateral ovarioles, oviducts, and a single genital canal [27]. Major morphological features of a typical eriophyid are shown in Fig. 8.1.

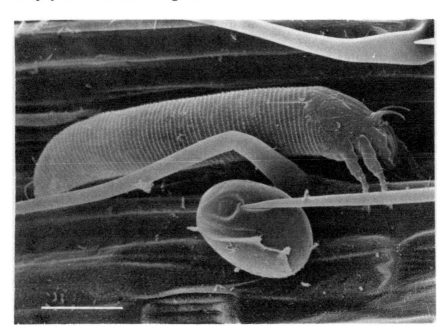

Fig. 8.1. Wheat curl mite *A. tulipae*. An egg and an adult mite on the upper surface of a wheat leaf. Bar = 50μ.

The eriophyids feed by sucking plant juices. *A. tulipae* probably only penetrate 5 μm into plant tissue, i.e., only into the epidermal cells, because of the structure and attachment of the stylets and rostrum [32]. In many cases, their stylets puncture plant tissue without causing apparent damage to the host. When severely affected, however, visible injuries may occur, resulting in leaf discoloration, malformations, galls, bud blasting, leaf puckering or rolling, and fasciation, that could be confused with those symptoms caused by viruses or other pathogens [9].

All eriophyids are phytophagous, many highly host-specific and having restricted host ranges. In extreme cases, they infect only certain varieties of one plant species. More commonly, many eriophyids are known to be pests of several plant species within one genus. Others, rare in number, have hosts in several genera. The eriophyids are incapable of sustaining themselves through long periods in the absence of a living host, thus requiring for their survival perennial plants or annuals that continuously grow in an overlapping manner throughout the year.

Experimental Mite Transmission

If specific mites are found to be involved in causing certain disease symptoms, (1) disease-free mites must be established on an appropriate healthy plant, (2) the mites must be given acquisition feeding, (3) the mites must be transferred to new test plants for inoculation feeding (Plates 8.1 and 8.2) and (4) the mites must be removed or killed and the inoculated plants observed for possible symptom development. Even if disease symptoms are observed, additional tests must be performed to ensure that the disease is caused by an infectious agent carried by the mites and not by a toxic substance released by the feeding mites. When disease symptoms continue to develop on mite-free plants, it can be taken as evidence that mite-induced diseases were caused by infectious agents. Nevertheless, this test cannot be entirely reliable because some toxins may have long-term effects, and some viruses are known to produce only mild, transitory symptoms in some cases. Furthermore, while transovarial transmission of plant viruses is not known for mites, a test of its possibility must be included [33].

Artificial transmission by alternative means is one of the effective methods to prove that an infectious agent is involved. Thus, transmission tests must be performed using sap extracted from diseased plants or, if that is not possible, using scions from diseased plants for grafting. In either test, precautions must be taken to ensure that all plants used remain mite-free.

DISEASE AGENTS TRANSMITTED BY ERIOPHYID MITES

Plant Diseases Associated with Eriophyid Mites

At present 16 diseases are either known to be associated or suspected to be associated strongly with Eriophyid mites as vectors of disease-causing (or associated) agents, which are essentially of three groups (Table 8.1). The first group consists of eight viruses with about 700 nm long flexuous particles. Most of them are distantly serologically related to each other. They resemble the potyviruses in inducing the formation of intracellular pinwheel inclusions, although there are no serological relationships to the true potyviruses by using antisera to virions. However, antiserum to the helper component of tobacco vein mottling virus, an aphid-transmitted potyvirus, reacted to a product of the cell-free translation of wheat streak mosaic virus (WSMV), which has an apparent Mr of 78 k, identical with that of the comparable products [34]. In spite of the absence of evidence for mite transmission, hordeum mosaic virus (HMV) and oat necrotic mottle virus are serologically related to WSMV, a mite-transmitted virus. Furthermore, HMV was shown to have some characteristics of both WSMV and Agropyron mosaic virus, a justification for the inclusion of the two viruses in this group. The second group consists of four diseases which are associated with the production of ovoid, double-membrane bodies (100–200 nm) in the cytoplasm of infected cells but which are absent in healthy tissues. However, none of these cases has been proven as the infectious agent. The third group consists of four fruit diseases which are associated with mites as known vectors but information on the causal agents is lacking or incomplete and merits further investigation.

Viruses with Flexuous Particles

Wheat streak mosaic virus (WSMV). Wheat streak mosaic was probably noted first in Nebraska as early as 1922 and since then it has caused serious losses in wheat in the central Great Plains and the western region of North America [29–32, 37, 38], Romania [36, 39], USSR [40, 41], Bulgaria [42], Turkey [43], Yugoslavia [44], and Jordan [36]. The virus causes a severe mosaic disease of most cultivars of winter wheat (Plates 8.3 and 8.4), oats, barley, rye, and some cultivars of maize [45]. It resembles potyviruses in morphology by being a flexuous rod-shaped virus (700 nm in length) and by inducing cylindrical or pinwheel-shaped cytoplasmic inclusions containing a 66 kDa protein [45]. Most natural isolates contain a capsid protein of 45 kDa and the type strain 47 kDa protein [46]. Using complementary DNA clones, prepared to the 3′

Table 8.1. Plant diseases associated with eriophyid mites as vectors of known and unknown etiological agents

Disease (mite)	Pathogen	Distribution	Transmitted by		Symptoms continue after mites killed	Noninfective mites from eggs	Key ref
			Sap	Graft			
Wheat streak mosaic (*Aceria tulipae*)	Flexuous particles (700 nm long)	N. America Middle East Europe	+	0	+	+	28,31, 35,45
Agropyron mosaic (*Abacarus hystrix*)	Flexuous particles (717 nm long)	N. America UK N. Europe	+	0	+	+	61,62, 65
Ryegrass mosaic (*A. hystrix*)	Flexuous particles (700 nm long)	N. America UK N. Europe	+	0	+	+	72,73
Hordeum mosaic (vector unknown)	Flexuous particles (698 nm long)	N. America	+	0	0	0	78,79
Oat necrotic mottle (vector unknown)	Flexuous particles (720 nm long)	Canada	+	0	0	0	81
Garlic mosaic (*A. tulipae*)	Flexuous particles (700 nm long)	Asia N. America Europe	+	0	+	+	86
Prunus latent (*Vasates fockeui*)	Flexuous particles (750 nm long)	Europe	+	0	0	0	(94) (95)

	Flexuous particles (725 nm long)						References
Spartina mottle (vector unknown)	DMB	UK	+	0	0	0	69
Wheat spot mosaic (*A. tulipae*)	DMB	N. America	−	0	+	+	102 / 104
Fig mosaic (*E. ficus*)	DMB	N. America / Middle East / Europe	−	+	+	+	116 / 117
Pigeon pea sterility (*E. cajani*)	DMB	India	−	0	+	0	(119) / (120)
Rose rosette (*P. fructiphilus*)	DMB	N. America	+	+	+	0	125 / 127
Currant reversion (*P. ribis*)	0	UK / Europe	−	+	+	0	134
Peach mosaic (*P. insidiosus*)	0	N. America / Europe	−	+	+	+	147 / 151 / 152
Cherry mottle leaf (*P. inaequalis*)	0	N. America / Europe / S. Africa	−	+	+	0	147 / 156
Wheat spotting (*A. mckenziei* and *E. tritici*)	0	USSR	+	0	0	0	(158)

+, tests with positive results; −, tests with negative results; 0, no tests or unsatisfactory tests; DMB, double-membrane body; (), to be confirmed.

portion of the WSMV genomic RNA, DNA sequence analysis revealed an open reading frame of 1,292 nucleotides, five potential sites for hydrolysis by a viral-coded protease, a relatively short (147 nucleotide) 3' untranslated region, and a polyadenylated (A = 16) 3' terminus [47]. Thus, as with potyviruses, the coat protein gene of WSMV is proximal to the 3' terminus of its genomic RNA. However, computer-based comparisons (Chou-Fasman prediction) of the structure of WSMV coat protein with those of known potyviruses showed only limited homology. *A. tulipae* is an efficient vector of WSMV and up to 60% of the mites in all stages of development, except the eggs, are viruliferous. Non-viruliferous mites are established by transferring the eggs to healthy wheat to hatch. A second vector species, *Aceria (Eriophyes) tosichella* Keifer, has been reported from Yugoslavia [44].

A. *tulipae* acquires WSMV during long feeding periods (15 min or longer) and retains it for long periods (up to nine days) including passage through molts. Only nymphs are capable of acquiring WSMV although both nymphs and adults can transmit it [29].

Wheat streak mosaic virus particles were found densely packed in the midgut and hindgut of *A. tulipae* reared on WSMV-infected wheat plants [48–50]. Since the gut contents which accumulated during prolonged feeding could be regurgitated, ingested virus particles might be recycled to the mouth parts to serve as inoculum. Virus-like particles were also found in the parenchymatous tissue in proximity to the intestine [49] or in the intestine [50] of *E. tulipae*. It was interpreted that virus transmission might occur through the intestinal wall into the parenchyma cells, then to the salivary glands, and into plant cells during feeding [50].

The mites and their eggs were capable of surviving the lowest temperatures that the varieties of winter wheat could survive when tested at sub-freezing temperatures of $-5\,^\circ$C to $-25\,^\circ$C for various periods of time [28]. Living plants are essential for the survival of mites and the virus. Thus, high disease incidence is encouraged by an overlapping succession of susceptible wheat grown nearby (Plates 8.5 and 8.6). The prevalence and severity of the disease are primarily dependent on the presence and pattern of wheat culture. Infected winter wheat carries the virus and mites through winter. The mites multiply rapidly during mild weather in the spring and summer and are spread by wind, carrying the virus to wheat and other susceptible plants in close proximity. The epidemiology of WSMV seems closely correlated with mite population dynamics [51], and with the grass host of the mites and the virus [52]. Maize may harbor infectious mites in immature ears until after the new winter wheat crops are sown, and thus serves as a source of infection, a frequent problem in parts of Ontario and Ohio [53–55].

Among many species tested for susceptibility to WSMV by sap inoculation or by mite transmission in North America, 15 perennial grass species have been found infected in the greenhouse and a few in the field [29, 57, 58]. However, none has been susceptible to the virus and at the same time capable of sustaining *E. tulipae*. In Europe, several perennials, such as *Boutelous curtipendula* (Michx.) Torr., *Festuca ovina* L., *Hordeum nodosum* L., and *Lolium perenne* L., have been reported to harbor WSMV in the field [42, 59]. It seems probable that some species of *Agropyron*, *Bouteloua*, *Elymus*, *Eragrostis*, and *Festuca*, or other grass species are likely both reservoirs of WSMV and supportive hosts of *E. tulipae* in North America [7]. The landing efficiency of *E. tulipae* is increased on wheat cultivars that have high densities of leaf trichomes. Thus, it was suggested that new cultivars developed for areas in which WSMV is a production constraint would probably benefit from having leaves with low trichome density [60].

Agropyron mosaic virus (AgMV). Agropyron mosaic was first found on quackgrass (*Agropyron repens*) in Virginia in 1934. The disease is common among grasses but rare on wheat, rye, and barley in North America [61–63], Britain [64], and Northern Europe [62]. The virus induces pale green or yellow mosaics, streaks, and dashes on leaf tissues but the symptoms are milder as plants reach maturity [62]. It has flexuous rods (710–717 nm long, 15 nm in diameter), and induces the production of intracellular pinwheel and tubular inclusions [62]. Two eriophyid mites, *Abacarus hystrix* and *Aculodes mckenziei*, are usually associated with AgMV on *Agropyron repens* or wheat in Ontario. In transmission tests, however, only *A. hystrix* was found to be the vector of AgMV with low transmission efficiency. Thus, only a few plants are likely to be infected in commercial wheat fields, even though the virus and mites can be perpetuated on an overlapping succession of wheat [65].

Spartina mottle virus (SMV). *Spartina anglica* C.E. Hubbard is a rhizomatous perennial grass that colonizes coastal mud flats in Britain and other countries [66], spreading over a total of 25,000 ha in nine countries [67]. *Spartina* marshes in Britain, Europe, and North America are used for grazing and silage production [68]. A virus distantly related to AgMV has been found in *Spartina* plants showing leaf mottling in England and Wales [69]. The flexuous particles measuring 725×12 nm were found in association with pinwheel inclusions and laminar aggregates. The virus is sap-transmissible to *S. anglica* but not transmitted by aphids or the eriophyid mite *A. hystrix* in the limited range of tests done. Its remote serological relationship to AgMV suggests that it may have a yet unknown mite vector.

Ryegrass mosaic virus (RMV). Only members of Gramineae are susceptible to RMV, which has been reported from Britain, Northern Europe [70, 71], and North America [72]. The virus is flexuous rods 700 nm long and 15 nm in diameter and causes light green to yellow mosaic on *Lolium* spp. and *Dactylis glomerata* L., natural hosts.

Among species of Eriophyidae found on mosaic-affected ryegrass, only *A. hystrix* transmitted RMV [73]. All instars of the mites from virus-infected plants transmitted the virus. Non-viruliferous mites produced from eggs acquired the virus in two hours in infected ryegrass and the percentage that became viruliferous increased with increasing acquisition feeding time up to 12 hr. Viruliferous mites lost their transmissibility within 24 hr on wheat immune to RMV.

Up to 30% losses in the herbage yield of Italian ryegrass (*L. multiflorum*) in southeastern England have been reported [74]. Although the spread of RMV transmitted by mites was rapid, its prevention was possible by applications of aldicarb [75]. There has been little information on mite resistance in ryegrasses [76], but a tolerant genotype is known to be twice as productive as a sensitive one [77].

Hordeum mosaic virus (HMV). Hordeum mosaic virus, reported from Alberta, Canada, causes a diffuse chlorotic mottle on barley leaves. No streaks develop except near the base of young leaves, in contrast to AgMV and WSMV, both of which cause streaks in infected barley and wheat leaves. The host range of HMV includes wheat, rye, and a number of other grass species. *A. tulipae* and *A. hystrix* did not transmit HMV [78]. However, the general similarities between HNV and the other mite-transmitted viruses point to a possibility that it too could be transmitted by a mite species that is yet to be discovered. No transmission tests have been carried out with mites found on naturally infected grasses in Alberta where the virus was originally found. Particles of HMV are flexuous rods 683–698 nm long. Disease symptoms are noted at 10–30°C. Most barley cultivars tested are moderately susceptible to HMV [79].

Oat necrotic mottle virus (ONMV). Oat necrotic mottle virus, flexuous rods of 720 nm × 11 nm, first described in 1966 [80], causes a stunting disease in oats and mild or symptomless infections on the genera *Poa, Bromus,* and *Lolium.* It has been reported only from Manitoba, Canada, but could be more widespread in the western plains of North America [81].

The symptoms on infected oats are narrow chlorotic lines and irregular mottle on emerging leaves followed by necrosis on the mottled leaves. Infected plants are moderately stunted. At 15–18°C mottling is more prominent, while at warmer temperatures necrosis appears sooner and is pronounced [81]. In greenhouse trials, ONMV caused a reduction in oat seed yield of 16 and 84% at 15 and 20°C respectively. It is not known how the virus is transmitted in nature, though the distant serological relationship [82] with WSMV suggests the possibility of mite transmission. The virus resembles RMV and WSMV in that RNA has a single-stranded, monopartite genome of 2.6 × 10⁶ Da [83]. Infected oat plants are usually found at the field margin, and symptomless *Poa compressa* and *P. canadensis* collected near cereal fields are frequently infected, an indication that the virus is harbored in the grasses [81].

Garlic mosaic virus (GMV). In 1979–80, 'tangle-top' of garlic became an important problem in all growing regions of the Philippines [84]. The country's garlic production was reduced by 40–50%. Examination of affected plants revealed that *A. tulipae* was associated with the disease. The tangling or twisting of leaves and large yellow blotches were caused by the mite's feeding activity [85], while short, pale yellow streaks on the lamina were due to a sap-transmissible virus [86]. The minimum acquisition and inoculation feeding times required by *A. tulipae* to transmit the virus were < 30 min each. Both the first and second instar nymphs of the mite were capable of virus acquisition but not the adult. Adult mites were able to transmit the virus only when they acquired it during their nymphal stages. The virus persisted in the mites for at least eight days and was not transmitted through eggs. The relationship of this virus [87] to GMV reported from Japan [88], Korea [89, 90], and elsewhere [91] has not been investigated and requires clarification. A mosaic disease affecting up to 32% of onion seed plants has been reported from the Ukraine. Transmission tests showed that *A. tulipae* mites from diseased plants induced the symptoms after being reared on healthy onion seedlings. Flexuous virus-like particles measuring 675 × 15 nm were detected in sap from infected onion plants and in ultrathin sections of mites from the diseased plants. The virus-like particles and *A. tulipae* were also found on garlic in the northern Caucasus [92]. However, relationships between GMV and a virus causing onion mosaic remain undetermined. Since *A. tulipae* from onions can colonize wheat with difficulty [93], it is worth investigating whether onion mosaic virus and GMV are capable of infecting wheat and other species of Gramineae.

Prunus latent virus (PLV). Virus infection of symptomless plums was found in Germany when the eriophyid mite *Vasates fockeui* (Nal.) was transferred from some plum trees suspected to be diseased to indicator plants, *Chenopodium foetidum*. The sap-transmissible virus was repeatedly transferred to other *C. foetidum* and *C. quinoa* plants, both of which produced necrotic lesions. Extracts of the tissue bearing local lesions contained flexuous particles 750 nm long [94].

Subsequent investigations showed that the symptoms on *Chenopodium* species could be caused by PLV transmitted by single mites that had fed on plum, and increases in mite population resulted in the increased infection rates [95]. Transmission occurred during inoculation feeding periods of 5 min and prolonged feeding periods up to 24 hr resulted in increases in transmission rates. The virus persisted in the mites for at least four days after feeding on diseased plums. There is no information regarding the establishment of non-viruliferous mites from eggs or virus-acquisition tests using non-viruliferous mites, or transmission of the virus recovered from local lesions on *Chenopodium* plants to plum seedlings.

Double-membrane Bodies (DMBs)

There are at least four diseases in which DMBs, 100–200 nm in diameter, have been found in the cytoplasm of diseased plants but not in comparative samples from healthy plants. These include wheat spot mosaic [96, 97], fig. [96, 98, 99], pigeon pea sterility [100], and rose rosette [101].

These bodies do not seem to contain ribosomes characteristic of bacteria, mycoplasma, and related organisms, nor an electron dense nucleoid characteristic of viruses (Figs. 8.2 and 8.3).

According to a recent study, a prominent feature of wheat spot mosaic-affected leaf cells is the proliferation of intracellular membranes, in particular endoplasmic reticulum (Figs. 8.4 and 8.5). The percentage of DMBs which showed a physical connection with endoplasmic reticulum was 14.4% of 333 DMBs examined. The percentage increased to 67.0% of 115 DMBs when serial sections of 8–15 planes were examined, suggesting that DMBs were originally formed from endoplasmic reticulum. In SEM observation, similar structures of DMBs were confirmed as oval-shaped bodies formed at the end of smooth endoplasmic reticulum [102]. In a separate cytological study, under conditions which allowed the lactoferrin-gold complex to bind to the nucleus and the electron dense areas within the nucleolus, DMBs were free of labeling, suggesting the absence of nucleic acids [103]. On the other hand the consistent occurrence of double-stranded (ds) RNAs in rose rosette-affected multiflora roses has been reported. However, the relationships between the dsRNAs and DMBs has not been determined.

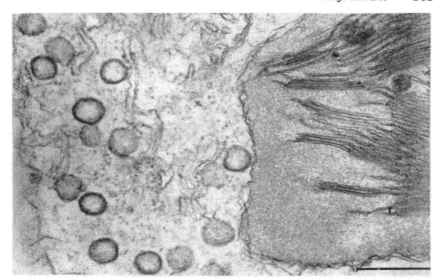

Fig. 8.2. Electron micrograph showing double-membrane bodies in the cytoplasm of a leaf of a *Triticum* × *Agropyron* hybrid (T-Ai) plant infected with wheat spot mosaic agent by mite transmission. Bar = 300 nm.

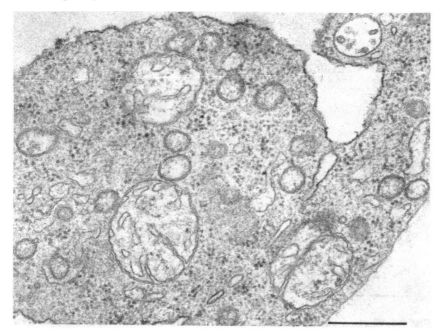

Fig. 8.3. Electron micrograph showing double-membrane bodies in the cytoplasm of a leaf of a *Triticum* × *Agropyron* hybrid (T-Ai) plant infected with wheat spot mosaic agent. Bar = 300 nm.

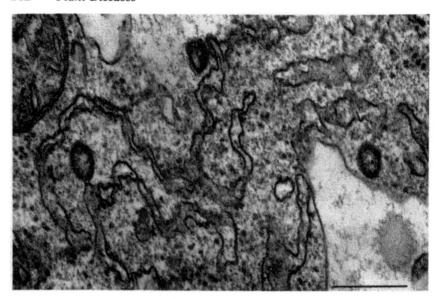

Fig. 8.4. Electron micrograph showing the proliferation of intracellular membranes, in particular endoplasmic reticulum in association with the double-membrane bodies in a hybrid wheat (T-Ai) leaf cell infected with wheat spot mosaic agent. Bar = 300 nm.

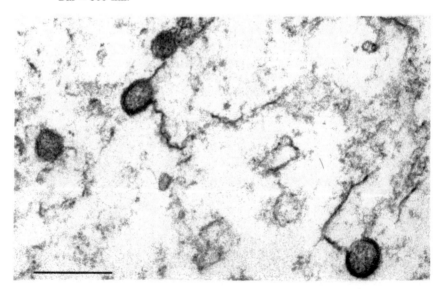

Fig. 8.5. Electron micrograph showing the close association of the double-membrane bodies with endoplasmic reticulum in a hybrid wheat (T-Ai) leaf cell infected with wheat spot mosaic agent. Bar = 300 nm.

Wheat spot mosaic agent. Wheat spot mosaic was found in Alberta in 1952 during mite transmission studies on WSMV [29, 104]. Some of the wheat seedlings inoculated by means of a single mite from naturally infected wheat developed symptoms dissimilar to those caused by WSMV, such as chlorotic spots, severe chlorosis, stunting, and necrosis, and the disease agent was not transmitted by sap inoculation.

The symptoms were at first attributed to the toxic effects of the mites, but they continued to develop even after the plants were freed of mites. When mites newly hatched from eggs were transferred to healthy wheat, they did not produce disease symptoms unless they had first fed on diseased wheat. All stages of the mites transferred from diseased plants induced similar symptoms. Different isolates of the disease agent caused symptoms that differed greatly in severity. Some can cause extreme leaf necrosis and death of emerging leaves, resulting in rapid death of whole infected plants. However, in these observations the effects of plant growth stage at the time of infection and of different environmental conditions were not defined and remain to be determined.

In the field the diseased plants were severely chlorotic and were also frequently infected with WSMV. The double infection appeared to induce synergistic effects on the severity of disease symptoms. When mites from a plant infected with both agents were tested singly, 65% transmitted wheat spot mosaic agent and 34% transmitted WSMV. Some mites transmitted both agents.

The host range of wheat spot mosaic agent determined by means of viruliferous mites was similar to but not identical with that of WSMV. It infected most of the common cereals tested including maize, and several wild annual grasses. Several *Triticum* × *Agropyron* hybrids immune to WSMV were susceptible to wheat spot mosaic agent.

A similar agent also transmitted by *A. tulipae* has been reported from Ohio [105] and North Dakota [106].

Fig mosaic agent. A mosaic of figs (*Ficus carica* L.) was first reported from California in 1933 [107]. Disease symptoms included mottling, spotting, and distortion of leaves, and occasionally leaf and fruit dropping. The distribution of the disease is worldwide, including Australia [108], Britain [109], Bulgaria [110], China [111], New Zealand [112], Italy [99], Libya [98], Yugoslavia [98], and India [113].

With regard to the identity of the disease-causing agent, it was initially concluded that mosaic symptoms of fig were caused by a presumptive virus because the symptoms appeared on new growth from symptomless seedlings grafted to diseased orchard trees [107]. However, such graft transmission experiments were made on mite-infested trees, and it was later pointed out that the possibility of the symptoms resulting from mite injury must be eliminated [114]. After elimination of mites with sulfur, graft transmission was demonstrated in the continued absence of mites [114]. Double-membrane bodies, measuring 120 to 160 nm in diameter, were found in the cytoplasm of fig mosaic-affected fig leaves from the United States [96, 100], Italy [99], Libya [98], and Yugoslavia [98]. Although they were not found in healthy leaf tissue, these structures have not been positively identified as infectious agents.

The host range of the fig mosaic agent is limited to members of the fig family, Moraceae. It affected most plants tested in more than 16 species of *Ficus* [113, 115]. *Cudrania tricuspidata* and *Morus indica* (mulberry) were also susceptible [116].

The fig mosaic agent could be transmitted from diseased to healthy seedlings of *F. carcia* by a single mite [117]. Both the minimum acquisition feeding and inoculation feeding periods were 5 min. Terminal buds were a better source of fig mosaic agent than laminae or petioles. The mites transmitted the fig mosaic agent a few hours after acquisition feeding and remained infectious for 6 to 10 days at room temperature. In California, all stages and both sexes of the mites were found throughout the year. Mites stayed in buds during the dormant season and were exposed as the buds opened in the spring. Then, eggs were laid on the stems and the leaves. In July mites moved from the leaves, entering the fruits [118].

Pigeon pea sterility agent. Pigeon pea (*Cajanus cajan* (L.) Millsp.), an important crop in India [119, 120] and elsewhere, is grown both as an annual crop and as a perennial. The disease is characterized by mosaic symptoms on foliage and partial sterility of the flowers. Initially it was believed that a virus transmitted by an eriophyid mite, *Aceria (Eriophyes) cajani* Channa Bosovanna, was involved as the pathogen [119, 120]. The causal agent was not transmitted by sap inoculation, dodder, leafhoppers, aphids, whiteflies, thrips, or two species of mites found on diseased pigeon pea plants. Symptoms developed in three to five weeks when 5–20 mites from diseased plants were placed on the plants. Mass transmission could be achieved by placing diseased, mite-infested leaves on the test plants. Establishment of non-viruliferous mites from eggs was not possible and tests for acquisition of the causal agent by non-viruliferous mites have not been performed.

Double-membrane bodies, about 100 to 200 nm in diameter, were found in the cytoplasm of leaf parenchyma cells of diseased plants but not in healthy controls [100].

Two resistant and 16 tolerant germplasm lines were selected among 234 tested for pigeon pea sterility under field conditions [121]. Since most cultivars are susceptible to the disease, these lines may offer potential sources for breeding programs for disease resistance.

Rose rosette agent. Rose rosette, also known as witches' broom of rose, is characterized by rosetting, mottling, and severe inflowering of many wild and cultivated species and cultivars of roses. The disease, described for the first time from Manitoba in 1941 [122] and from California in 1953 [123], has been found in the West and Midwest in North America and is spreading eastward [124]. The disease agent is easily transmitted by grafting [125, 126] and by sap [126] or mechanical inoculation [127]. Recently Di et al. (1990) [127] have reported the association with rose rosette of four double-stranded RNAs of Mr 2.9, 1.2, 1.0 and 0.93×10^6, which were consistently found with the causal agent by grafting and by mechanical transmission to multiflora plants [120]. However, it was not possible to demonstrate that the RNAs associated with the rosette disease are associated with the etiologic agent of the disease and not with a plant response to the agent [127]. A host range study using graft inoculation indicated that none of the eight members of the Rosaceae tested, except the genus *Rosa*, was susceptible to the rose rosette agent [125]. No evidence was obtained for transmission of the causal agent through seed or fruit harvested from infected plants [125].

The disease agent is transmitted from plant to plant by an eriophyid mite, *Phyllocoptes fructiphilus* Keifer, which is usually found on diseased rose bushes and wild roses which serve as a carryover host [128, 129]. Double-membrane bodies, 120–150 nm in diameter, were found in the cytoplasm of parenchyma, phloem, and epidermal cells of infected roses [101]. The particles from diseased rose plants were reportedly transmitted to healthy multiflora rose plants in the greenhouse using *P. fructiphilus* as a vector [101], but experimental data were not presented. No tests were reported to determine if rosette symptoms induced by the mites continued even after all mites were removed, nor to determine the relationship between the RNAs and DMBs. Also control experiments are required using non-viruliferous isolates of the mites established on healthy plants.

Under field conditions in West Virginia, 17 to 30 days after mite transmission, the inoculated plants of multiflora rose developed a bright red to dark red mosaic pattern on new leaves. One to three months later, numerous lateral shoots started developing on the infected plants. A proliferation of these shoots forms the symptom known as witches' broom. In early spring, infected plants produced numerous red lateral shoots and thick clusters of reddish green leaves which were often smaller than normal leaves [126]. Regardless of the rate or method of application, oxytetracycline was ineffective in causing a remission of the rosette symptoms in multiflora roses [125].

Disease Agents of Unknown Characteristics

Currant reversion agent. The disease of black currant (*Ribes nigrum* L.) known as reversion is widespread in Britain and probably throughout Europe and is characterized by the foliage appearance which resembles reversion to the wild type because of the modified leaf morphology. Affected bushes consequently cease to bear fruit. In reality, however, it is the symptom caused by the disease-causing agent that is transmitted by the eriophyid mite, *Cecidophyopsis (Phytoptus / Eriophyes) ribis* (Westw.). While the involvement of mites as the vector was suggested as early as the 1920s [130, 131] and experimentally demonstrated [132, 133, 134], transmission of the disease agent using single mites was successfully established [135], and later independently confirmed [133]. Although the disease agent was suggested to be mycoplasma [136–138], the evidence for it is not conclusive and the nature of the pathogen still needs to be established. The mites transmitted the reversion agent during all except the first two weeks of their active migration period throughout April to June [139]. A correlation could not be established between the development of mite infestations and currant reversion on bushes in a young plantation. An interesting observation is that reverted bushes were more susceptible to the mite infestation than healthy bushes. The young buds of bushes affected with reversion by graft transmission were over 100-fold more susceptible to the mites than comparable healthy bushes. This difference was interpreted to mean that the disease resulted in the increased vulnerability and accessibility of apical and axillary buds as the result of reduction in the density of hairs on the young tissues, since the highest infestations of mites were recorded on reverted bushes, whereas mite populations were low when reversion was absent on bushes [140]. However, such a difference may also be caused by physiological changes following infection [95]. A recent physiological study has indicated that levels of free endogenous gibberellins in healthy black currants were considerably higher than those in reversion-affected plants. In contrast, gibberellin inhibitor levels were consistently higher in diseased plants in comparison with the healthy ones [141].

Isolation of a virus identical with PVY has been reported from reversion-diseased black currants [142, 143]. It was claimed that the virus was transmitted by both nymphs and adults of gall mites (*C. ribis*), but not through eggs. The mites acquired the virus in three hours and it persisted for 25 days. The highest transmission was achieved after 48 hr acquisition feeding. Black currants were re-infected by mechanical inoculation with the virus in sap from *Nicotiana glutinosa*. Since there is no previous confirmed record of a virus being vectored both by aphids and by eriophyid mites, it is important to ascertain whether it represents a case of dual infections with a mite-transmitted virus and an aphid-transmitted virus.

Peach mosaic agent. Peach mosaic was found first in Texas and Colorado in 1931 and has subsequently been recognized in many fruit-growing areas in the United States [144–146] and Europe [147]. The causal agent is bud- or graft-transmissible to peach, plum, apricot, and almond [147]. The relationship between peach mosaic and the pathogen of peach latent mosaic described in France is still unclear [148, 149]. The symptoms on peach include a mosaic pattern on leaves, and the infected leaves remain small, narrow, crinkled, and irregular in shape. Internodes become shortened and a profuse growth of leaf axil buds is noted. Flower symptoms are most conspicuous on cultivars with pink blossoms where an unusual color-break often develops on the petals, resulting in irregular streaks, white spots, and line patterns. The petals are often crinkled and dwarfed, and on affected trees the number of emerging buds increases markedly [147]. Retarded short growth and foliation result in a stubby and rosetted appearance.

The disease is of high economic importance, and seems to spread very quickly, causing heavy losses to the fruit industry. In Colorado only seven diseased trees were found in 1931, whereas in 1935 30,467 were found to be affected [150]. In a tree-removal program in California, 204,193 diseased peach trees were destroyed. Removal of infected trees, nursery inspection, and quarantine procedures have been effective in limiting the spread of the disease [150].

The causal agent of peach mosaic is transmitted by an eriophyid mite, *Phytoptus (Eriophyes) insidiosus* (Wilson and Keifer) [151, 152]. In transmission tests using a single mite, 2.5% of *P. insidiosus* transmitted peach mosaic. With 30–50 mites per test plant, 67–75% of the test seedlings developed symptoms. The mites retained infectivity for at least two days in the absence of a virus source. Non-viruliferous mites were obtained from eggs after hatching on healthy seedlings. Control plants with *P. insidiosus* mites from healthy peach did not develop symptoms under comparable conditions. The properties of the causal agent of peach mosaic remain unknown. Chemotherapy has been attempted without positive results. The peach mosaic agent was not inactivated by heat treatment. Effective control of the spread of peach mosaic was achieved by chemical spray targetted at mites in peach orchards where the disease is heavy [153].

Cherry mottle leaf agent. Cherry mottle leaf disease was recorded in Oregon, USA as early as 1930 [154, 155]. It is considered to be one of the most damaging diseases of sweet cherry in some areas. It has been reported from North America [156], Europe [147], and South Africa [147].

The most characteristic symptoms of cherry mottle leaf are the irregular chlorotic mottle and distortion on leaves, especially on the terminal ones. The puckering, tattering, shot-hole, and reduction in leaf size vary according to host cultivar and disease severity. The fruits of severely affected trees are small and flavorless and their ripening is delayed but no deformations appear. Infected trees appear rosetted due to inhibition of terminal growth and shortening of internodes [147]. The disease is most prevalent in orchards near wild bitter cherry (*Prunus emarginata*) [152].

The disease agent was transmitted by budding, grafting, and chip budding, as well as by the leaf mite *Phytoptus (Eriophyes) inaequalis*, a parasite of *Prunus emarginata* (Doublas) which was found to be the principle host of the mite and the disease agent [157]. The causal agent was not transmitted by seed or mechanically and its properties have not been determined.

Wheat spotting agent. The disease was observed on winter wheat, *Lolium perenne*, and a meadow fescue (*Festuca pratensis*) in the Soviet Far East [158]. The disease agent was sap-transmissible but was not transmitted by seeds, soil, or plant debris. The principal symptoms are light green, yellowish green, or white spotting on the stems and leaves. The mites *Aculodes mckenziei* and *Aceria (Eriophyes) tritici* were reported to be effective vectors. In inoculation tests, *A. mckenziei* transmitted the disease agent to wheat, barley, *Agrostis alba, Phelum arundinacea*, and *L. perenne*, and to 27 *Bromus, Lolium, Sorghum*, and *Calamgrostis* spp. The normal incubation period in plants was two to seven days, and sometimes 10 days or longer. Other properties of the disease agent remain unknown.

CONCLUSIONS

Slykhuis (1965) [4] cited several reports of suspected virus transmission by tetranychids. However, in most cases little or no experimental evidence in support of these cases has been provided. In spite of two additional, independent studies on the vector relationships between *T. urticae* and PVY, the initial claim [10] of spider mites being the vector of PVY has not been confirmed.

Since Slykhuis' (1973) [6] and Oldfield's (1970) [8] last reviews on mite transmission of plant viruses, Tenuipalpidae and Tarsonemidae have been reported as the suspected vectors of two new viruses, a polyhedral virus 35 nm in diameter and a rhabdovirus-like virus with particles 100–150 × 40 nm. Although an orchid virus morphologically resembling the latter virus has been described [20], the association of vector mites with the virus has not been found yet, and requires further investigation.

More information on specific vector relationships between eriophyid mites and various plant disease agents has been accumulated. The available evidence indicates that three distinct groups exist for plant diseases associated with eriophyid mites as vectors. The first group consisting of eight diseases is the one associated with potyvirus-like particles represented by WSMV. However, unlike true potyviruses in which capsid protein and helper protein are involved in transmission by aphids [159], the molecular basis of transmission by the mites has not been elucidated for eriophyid-borne viruses and requires intensive studies. The second group consists of four diseases, wheat spot mosaic, fig mosaic, pigeon pea sterility, and rose rosette, all of which are associated with specific eriophyid mites as vectors and have DMBs of 100 to 200 nm in diameter as suspected causal agents. At present the identity of the DMBs as an etiological agent has not been established. Detailed molecular characterization of the DMBs is expected to yield useful information for our understanding of their etiology. The third group includes four diseases, currant reversion, peach mosaic, cherry mottle leaf, and wheat spotting, whose causal agents remain uncharacterized, although the diseases are known to be associated with specific mite vectors. As more information on molecular aspects of the causal agents becomes available, some of them will likely be recategorized to either the first or the second group.

ACKNOWLEDGEMENTS

This work was supported in part by a grant from the Western Grains Research Foundation. Thanks are due to G. Oldfield, R. Conner, and V. Muniyappa for supplying specimens, and to E.E. Lindguist for useful comments, and to M.H. Chen, S.T. Ohki, K.S. Zaychuk, T. Tribe, and G. Figueiredo for assistance during the course of this study.

REFERENCES

1. Amos, J., R.G. Hatton, R.C. Knight and A.M. Massee. Experiments in the transmission of reversion in black currants. *Ann. Rep. East Malling Res. Sta. Kent*, Suppl. II, 126, 1927.
2. Slykhuis, J.T. Mite transmission of plant viruses. In: *Biological Transmission of Disease Agents*. K. Maramorosch, ed. New York: Academic Press, p. 41, 1962.
3. Slykhuis, J.T. Mite transmission of plant viruses. *Adv. Acarol.*, 1, 326, 1963.
4. Slykhuis, J.T. Mite transmission of plant viruses. *Adv. Virus Res.* 11, 97, 1965.

150   Plant Diseases

5. Slykhuis, J.T. Mites as vectors of plant viruses. In: *Viruses, Vectors, and Vegetation.*
    K. Maramorosch, ed. New York: Interscience, p. 121, 1969.
6. Slykhuis, J.T. Viruses and mites. In: *Viruses and Invertebrates.* A.J. Gibbs, ed.
    Amsterdam: North-Holland, p. 391, 1973.
7. Slykhuis, J.T. Mites. In: *Vectors of Plant Pathogens.* K. Harris and K. Maramorosch,
    eds. New York: Academic Press, Chap. 14, 1980.
8. Oldfield, G.N. Mite transmission of plant viruses. *Ann. Rev. Entomol.*, 15, 343, 1970.
9. Jeppson, L.R., H.H. Keifer and E.W. Baker. *Mites Injurious to Economic Plants.*
    Berkeley: University of California Press, 1975.
10. Schulz, J.T. *Tetranychus telarius* (L.) new vector of Virus Y. *Plant Dis. Rep.*, 47, 594,
    1963.
11. Fritzsche, R., K. Schmelzer and H. Schmidt. Evaluation of the ability of *Tetranychus
    urticae* Koch as a vector of plant viruses. *Arch. Pfl.-Schutz*, 3, 89, 1967.
12. Orlob, G.B. Relationships between *Tetranychus urticae* Koch and some plant viruses.
    Virology, 35, 121, 1968.
13. Hollings, M. and A.A. Brunt. Potyviruses. In: *Handbook of Plant Virus Infections.*
    E. Kurstak, ed. Amsterdam: Elsevier/North Holland, p. 731, 1981.
14. Francki, R.I.B., R.G. Milne and T. Hatta. *Atlas of Plant Viruses*, Vol. 1. Boca Raton,
    Florida: CRC, 1985.
15. Knorr, L.C. and E.P. Ducharme. The relationship between Argentina's lepra explosiva
    and Florida's scaly bark, with implications for the Florida citrus grower. *Plant Dis.
    Rep.*, 35, 70, 1951.
16. Knorr, L.C., H.A. Denmark and H.C. Burnett. Occurrence of Brevipalpus mite,
    leprosis, and false leprosis on citrus in Florida. *Fla. Entomol.*, 51, 11, 1968.
17. Knorr, L.C. Studies on the etiology of leprosis in citrus. In: *Proc. 4th Confl. Int. Org.
    Citrus Virol.* J.F.L. Childs, ed. Gainesville: Univ. Fla. Press, p. 332, 1968.
18. Rosettie, V., C.C. Lasca and S. Negretti. New developments regarding leprosis and
    zonate chlorosis of citrus. *Proc. 1st Int. Citrus Symp.* 3, 1453, 1969.
19. Kitajima, E.W., G.W. Muller, A.S. Costa and W. Yuki. Short, rod-like particles
    associated with Citrus leprosis. *Virology*, 50, 254, 1972.
20. Doi, Y., M.U. Chang and K. Yora. Orchid fleck virus. *C.M.I./A.A.B. Descriptions of
    Plant Viruses*, No. 183, 1977.
21. Duvel, D. and K.R. Peters. Virusähnliche Partikeln in *Dendrobium antennatum* Ldl.
    Gartenwelt, 71, 52, 1971.
22. Petzold, H. Der elektronenmikroskopische Nachweis eines bacilliformen Virus an
    blattfleckenkranken Dendrobien. *Phytopathol. Z.*, 70, 43, 1971.
23. Lesemann, D. and J. Begtrup. Elektronenmikroskopischer Nachweis eines
    bacilliformen Virus in Phalaenopsis. *Phytopathol. Z.*, 71, 257, 1971.
24. Kitajima, E.W., A. Blumenschein and A.S. Costa. Rodlike particles associated with
    ringspot symptoms in several orchid species in Brazil. *Phytopathol. Z.*, 81, 280. 1974.
25. Lesemann, D., and S. Doraiswamy. Bullet-shaped virus-like particles in chlorotic and
    necrotic leaf lesions of orchids. *Phytopathol. Z.*, 83, 27, 1975.
26. Shikata, E., S. Kawano, T. Senboku, E.R. Tiongco and K. Miyajima. Small virus-like
    particles isolated from the leaf sheath tissues of rice plants and from the rice
    tarsonemid mites, *Steneotarsonemus spinki* Smiley (Acarina, Tarsonemidae). *Ann.
    Phytopathol. Soc. Jpn.*, 50, 368, 1984.
27. Whitmoyer, R.E., L.R. Nault and O.E. Bradfute. Fine structure of *Aceria tulipae*
    (Acarina: Eriophidae). *Ann. Entomol. Soc. Am.*, 65, 201, 1972.
28. Keifer, H.H., E.W. Baker, T. Kono, M. Delfinado and W.E. Styer. An illustrated guide
    to plant abnormalities caused by eriophyid mites in North America. USDA, Agriculture
    Handbook No. 573, 178 pp., 1982.

29. Slykhuis, J.T. *Aceria tulipae* Keifer (Acarina: Eriophyidae) in relation to the spread of wheat streak mosaic. *Phytopathology*, 45, 116, 1955.

30. Staples, R. and W.B. Allington. Streak mosaic of wheat in Nebraska and its control. Univ. Nebraska Agr. Exp. Sta. Res. Bull. 178, 41 pp., 1956.

31. Nault, L.R. and W.E. Styler. The dispersal of *Aceria tulipae* and three other grass-infesting eriophyid mites in Ohio. *Ann. Entomol. Soc. Am.*, 62, 1446, 1969.

32. Orlob, B.G. Feeding and transmission characteristics of *Aceria tulipae* Keifer as vector of wheat streak mosaic virus. *Phytopathol. Z.*, 55, 218, 1966.

33. Slykhuis, J.T. Methods for experimenting with mite transmission of plant viruses. In: *Methods in Virology*. K. Maramorosch, ed., New York: Academic Press, Chap. 10, 1967.

34. Hiebert, E., D.W. Thornberry and T.P. Pirone. Immunoprecipitation analyses of potyviral *in vitro* translation products using antisera to helper component of tobacco vein mottling virus and potato virus Y. *Virology*, 1235, 1, 1984.

35. Slykhuis, J.T. Wheat streak mosaic in Alberta and factors related to its spread. *Can. J. Agr. Sci.*, 33, 195, 1953.

36. Slykhuis, J.T. and W. Bell. New evidence on the distribution of wheat streak mosaic virus and the relation of isolates from Rumania, Jordan and Canada. *Phytopathology*, 53, 236, 1963.

37. Ashworth, L.J., Jr. and M.C. Futrell. Sources, transmission, symptomatology and distribution of wheat streak mosaic virus in Texas. *Plant Dis. Rep.*, 45, 220, 1961.

38. McKinney, H.H., M.K. Brakke, E.M. Ball and R. Staples. Wheat streak mosaic virus in the Ohio Valley. *Plant Dis. Rep.*, 50, 951, 1966.

39. Pop, I. Die Strichelvirose des Weizens in der Rumanischen Volksrepublik. *Phytopathol. Z.*, 43, 325, 1962.

40. Razvyazkina, G.M., E.A. Kopkova and U.V. Belyanchikova. Wheat streak mosaic virus. *Zashchita Rast. Moskva*, 8, 54, 1963.

41. Stein-Margolina, V.A., N.E. Cherin and G.M. Razvyazkina. The wheat streak mosaic virus in plant cells and the mite vector. *Dokl. Akad. Nauk SSR*, 169, 1446, 1966.

42. Markov, M.P., P. Kaitazova and I. Stefanov. Identification of wheat streak mosaic virus in Bulgaria. '*Rasteniev' dni Nauki*, 12, 130, 1975.

43. Bremer, K. Comparison of four virus isolates of wheat streak mosaic from Turkey. *Phytopathol. Med.*, 12, 67, 1973.

44. Tosic, M. Virus diseases of wheat in Serbia. *Phytopathol. Z.*, 70, 145, 1971.

45. Brakke, M.K. Wheat streak mosaic virus. *C.M.I./A.A.B. Descriptions of Plant viruses*, No. 48. 1971.

46. Brakke, M.K., R.N. Skopp and L.C. Lane. Degradation of wheat streak mosaic virus capsid protein during leaf senescence. *Phytopathology*, 80, 1401, 1990.

47. Niblett, C.L., L.A. Calvert, D.M. Stark, S.A. Lommel and R.N. Beachy. Characterization of the coat protein gene of wheat streak mosaic virus. *Phytopathology*, 78, 1561, 1988.

48. Paliwal, Y.C., and J.T. Slykhuis. Localization of wheat streak mosaic virus in the alimentary canal of its vector *Aceria tulipae* Keifer. *Virology*, 32, 344, 1967.

49. Paliwal, Y.C. Fate of plant viruses in mite vectors and nonvectors. In: *Vectors of Plant Pathogens*. K.F. Harris and K. Maramorosch, eds. New York: Academic Press, Chap. 15, 1980.

50. Takahashi, Y. and G. Orlob. Distribution of wheat streak mosaic virus-like particles in *Aceria tulipae*. *Virology*, 38, 230, 1969.

51. Stein-Margolina, V.A., N.E. Cherni and G.M. Razvyazkina. Phytopathogenic viruses in cells of the plant and mite vector. *IZV. Akada. Nauk. SSR, Ser. Biol.*, 62, 1969.

52. Lamey, H.A. and R.G. Timian. Wheat streak mosaic. North Dakota State Univ. Coop. Ext. Serv. Circ., 640 pp., 1979.

53. Somsen, H.W. and W.H. Sill, Jr. The wheat curl mite, *Aceria tulipae* Keifer, in relation to epidemiology and control of wheat streak mosaic. Kan. Agric. Exp. Sta. Res. Publ. 162, 4 pp. 1970.

54. Williams, L.E., D.T. Gordon, L.R. Nault, L.J. Alexander, O.E. Bradfute and W.R. Findley. A virus of corn and small grains in Ohio and its relation to wheat streak mosaic virus. *Plant Dis. Rep.*, 51, 207, 1967.

55. Paliwal, Y.C., J.T. Slykhuis and R.E. Wall. Wheat streak mosaic virus in corn in Ontario. *Can. Plant Dis. Surv.*, 46, 8, 1966.

56. Gates, L.F. The potential of corn and wheat to perpetuate wheat streak mosaic in southern Ontario. *Can. Plant Dis. Surv.*, 50, 59, 1970.

57. Sill, W.H., Jr. and R.V. Connin. Summary of the known host range of the wheat streak mosaic virus. *Trans. Kansas Acad. Sci.*, 56, 411, 1953.

58. Connin, R.V. The host range of the wheat curl mite, vector of wheat streak mosaic. *J. Econ. Entomol.*, 49, 1, 1956.

59. Razvyazkina, G.M., E.A. Kapkova and Y.V. Belyanchikova. Wheat striate mosaic virus. *Azshch. Rast. Moskva*, 8, 54, 1963.

60. Harvey, T.L., T.J. Martin and D.L. Seifers. Wheat curl mite and wheat streak mosaic in moderate trichome density wheat cultivars. *Crop Sci.*, 30, 534, 1990.

61. Slykhuis, J.T. *Agropyron* mosaic as a disease of wheat in Canada. *Can. J. Bot.*, 40, 1439, 1962.

62. Slykhuis, J.T. Agropyron mosaic virus. *C.M.I. /A.A.B. Descriptions of Plant Viruses.* No. 118, 1973.

63. Shepard, J.F. Occurrence of Agropyron mosaic virus in Montana. *Plant Dis. Rep.*, 52, 139, 1968.

64. Catherall, P.L. and J.A. Chamberlain. Occurrence of agropyron mosaic virus in Britain. *Plant Pathol.*, 24, 155, 1975.

65. Slykhuis, J.T. 1969. Transmission of agropyron mosaic virus by the eriophyid mite *Abacarus hystrix. Phytopathology*, 59, 29, 1969.

66. Hubbard, J.C.E. *Grasses.* 2nd Edn., Harmondsworth, Middlesex, England: Penguin Books, pp.354–361, 1968.

67. Ranwell, D.S. World resources of *Spartina townsendii* (sensu lato) and economic use of *Spartina* marshland. *J. Appl. Ecol.*, 4, 239, 1967.

68. Hubbard, J.C.E. and D.S. Ranwell. Cropping of *Spartina* salt marsh for silage. *J. Brit. Grassland Soc.*, 21, 214, 1966.

69. Jones, P. Leaf mottling of *Spartina* species caused by a newly recognized virus, spartina mottle virus. *Ann. Appl. Biol.*, 94, 77, 1980.

70. Beemster, A.B.R. Ryegrass mosaic virus in the Netherlands. *Meded. Rijksfac. Landbouwwet.*, 28, 749, 1976.

71. Pfeffer, B. and H. Pfeffer. Schaderreger an *Lolium*-Arten, ihre wirtschaftliche Bedeutung sowie Bekampfungsmoglichkeiten. *Nachrichtenbl. Pfl. Schutz.*, 39, 221, 1985.

72. Slykhuis, J.T. and Y.C. Paliwal. Ryegrass mosaic virus. *C.M.I. /A.A.B. Descriptions of Plant Viruses* No. 86. 4 pp., 1972.

73. Mulligan, T. The transmission by mites, host range and properties of ryegrass mosaic virus. *Ann. Appl. Biol.*, 48, 575–579, 1960.

74. Heard, A.J. and E.T. Roberts. Disorders of temporary ryegrass swards in southeast England. *Ann. Appl. Biol.*, 81, 240, 1975.

75. Gibson, R.W. Rapid spread by mites of ryegrass mosaic virus from old sward to seedling ryegrass and its prevention by aldicarb. *Plant Pathol.*, 30, 25, 1981.

76. Gibson, R.W. and A.J. Heard. Selection of perennial and Italian ryegrass plants resistant to ryegrass mosaic virus. *Ann. appl. Biol.*, 84, 429, 1976.

77. Wilkins, P.W. and D.H. Kides. Tolerance to ryegrass mosaic virus, its assessment and effect on yield. *Ann. Appl. Biol.*, 83, 399, 1976.

78. Slykhuis, J.T. and W. Bell. Differentiation of agropyron mosaic, wheat streak mosaic, and a hitherto unrecognized hordeum mosaic virus in Canada. *Can. J. Bot.*, 44, 1191, 1966.

79. Slykhuis, J.T. Hordeum mosaic. In: *Compendium of Barley Diseases*. D.E. Mathre, ed., St. Paul, Minnesota: Amer. Phytopathol. Soc., p. 52, 1982.

80. Gill, C.C. and P.H. Westdal. Virus diseases of cereals, and vector populations in the Canadian Prairies during 1965. *Can. Plant Dis. Surv.*, 46, 18, 1966.

81. Gill, C.C. Oat necrotic mottle virus. *C.M.I./A.A.B. Descriptions of Plant Viruses*, No. 169, 4 pp., 1976.

82. Gill, C.C. Serological properties of oat necrotic mottle virus. *Phytopathology*, 66, 415, 1976.

83. Gill, C.C. Some properties of the protein and nucleic acid of oat necrotic mottle virus. *Can. J. Plant Pathol.*, 2, 86, 1980.

84. Ahmed, K.M. and D.A. Benigno. Investigation into the relationship of the eriophyid mite (*Aceria tulipae* Keifer) with the 'tangle-top' and mosaic disease of garlic. *Bangladesh J. Agr. Res.*, 9, 38, 1984.

85. Smally, E.B. The production on garlic by an eriophyid mite of symptoms like those produced by viruses. *Phytopathology*, 46, 346, 1956.

86. Ahmed, K.M. and D.A. Benigno. Virus-vector relationship in mosaic disease of garlic. *Ind. Phytopathol.*, 38, 121, 1985.

87. Ahmed, K.M. and D.A. Benigno. Purification of garlic mosaic virus. *Philippine Agriculturalist*, 69, 193, 1986.

88. Sako, N. and N. Nagao. A virus disease of garlic. II. Fine structure of garlic mosaic leaf tissues. *Ann. Phytopathol. Soc. Jpn.*, 43, 114, 1978.

89. La, Y.J. Studies on garlic mosaic virus, its isolation, symptom expression in test plants, physical properties, purification, serology and electron microscopy. *Korean J. Plant Protect.*, 12, 93, 1973.

90. Chang, M.U., W.W. Park, J.D. Chung, K.B. Lim and Y.J. La. Distribution of garlic latent virus and garlic mosaic virus infected garlic tissues. *J. Korean Soc. Hort. Sci.*, 29, 253, 1990.

91. Shulman, N.I., T.D. Verderevskaya, E.S. Demidov, O.I. Kosakovskaya and O.O. Tamina. Serological diagnosis of garlic mosaic virus. *Izv. Akad. Nauk Mold. SSR, Ser. Biol. Khim. Nauk*, 67, 1989.

92. Razvyazkina, G.M. Das Zwiebelmosaikvirus and seine Verbreitung im Freiland. *Tagungsbericht Deutsche Akademie der Landwirtschaftswissenschaften zu Berlin.* 115, 69, 1971.

93. Del Rosario, M.S. and W.H. Sill, Jr. Physiological strains of *Aceria tulipae* and their relationships to the transmission of wheat streak mosaic virus. *Phytopathology*, 55, 1154, 1965.

94. Proeseler, G. and H. Kegler. Übertragung eines latenten Virus von Pflaume durch Gallmilben (Eriophyidae). *Monatsberichte Deut. Akad. Wis. Berlin*, 8, 472, 1966.

95. Proeseler, G. Die Gallmilbe *Cecidophyopsis ribis* (Westw.) als Schadling der Johannisbeeren. *Arch. Phytopathol. Pfl. Schutz*, 9, 383, 1973.

96. Bradfute, O.E., R.E. Whitmoyer and L.R. Nault. Ultrastructure of plant leaf tissue infected with mite-borne viral-like pathogens. *Proc. Elec. Microsc. Soc. Am.*, 28, 178, 1980.

97. Hiruki, C., M.H. Chen and S.T. Ohki. Vesicular bodies associated with wheat spot mosaic, semipersistently transmitted by wheat curl mite. *5th Internat. Cong. Plant Pathol.* Abst. St 1-2-4, 1988.

98. Plavsic, B. and D. Milicic. Intracellular changes in trees infected with fig mosaic. *Acta Hort.*, 110, 281, 1980.

99. Appiano, A. Cytological observations on leaves of fig infected with fig mosaic. *Caryologia*, 35, 152, 1982.

100. Hiruki, C., unpublished data, 1989.

101. Gergerich, R.C. and K.S. Kim. A description of the causal agent of rose rosette disease. *Ark. Farm Res.* 32, 7, 1983.

102. Chen, M.H. and C. Hiruki. The ultrastructure of the double membrane-bound bodies and endoplasmic reticulum in serial sections of wheat spot mosaic-affected wheat plants. *Proc. XII Internat. Cong. Electr. Microsc. 1990*, 3, 694, 1990.

103. Zaychuk, K.S., M.H. Chen and C. Hiruki. Protein-gold labelling of ultrathin sections of wheat spot mosaic-affected wheat plants. *Proc. XII Internat. Cong. Electr. Microsc.* 3, 680, 1990.

104. Slykhuis, J.T. Wheat spot mosaic, caused by a mite-transmitted virus associated with wheat streak-mosaic. *Phytopathology*, 46, 682, 1956.

105. Nault, L.R. and W.E. Styr. Transmission of an eriophyid-borne wheat pathogen by *Aceria tulipae. Phytopathology*, 60, 1616, 1970.

106. Edwards, M.C. and M.P. McMillan. A newly discovered wheat disease of unknown etiology in eastern North Dakota. *Plant Dis.* 72, 362, 1988.

107. Condit, I.J. and W.T. Horne. A mosaic of the fig in California. *Phytopathology*, 23, 887, 1933.

108. Noble, R.J. Australia: summary of plant diseases recorded in New South Wales for the season 1932–33. *Internat. Bull. Plant Prot.*, 8, 3, 1934.

109. Ainsworth, G.C. Fig mosaic. *J. Roy. Hort. Soc.*, 9, 532, 1935.

110. Atanasoff, D. Old and new virus diseases of trees and shrubs. *Phytopathol. Z.*, 8, 197, 1935.

111. Ho, W.T.H. and L.Y. Li. Preliminary notes on the virus diseases of some economic plants in Kwangtung Province. *Lingnan Sci. J.* 15, 67, 1936.

112. Li, L.Y. and C.H. Procter. A virus disease of fig in New Zealand. *N.Z.J. Sci. Tech.*, A, 26, 88, 1944.

113. Burnett, H.C. Additional hosts of the fig mosaic virus. *Plant Dis. Rep.*, 46, 693, 1962.

114. Nagaich, B.B. and K.S. Vashisth. Mosaic of *Ficus* spp. in India. *Curr. Sci.*, 31, 166, 1962.

115. Vashisth, K.S. and B.B. Nagaich. *Morus indica*—an additional host of fig mosaic. *Ind. Phytopathol.*, 18, 315, 1965.

116. Flock, R.A. and J.M. Wallace. Transmission of fig mosaic by the eriophyid mite *Aceria ficus. Phytopathology*, 45, 52, 1955.

117. Proeseler, G. Beziehungen zwischen Virus, Vektor und Wirtspflanze am Beispiel des Feigen-Mosaik-Virus und *Aceria ficus* Cotte (Eriophyidea). *Acta Phytopathol. Acad. Sci. Hung.*, 7, 179, 1972.

118. Baker, E.W. The fig mite, *Eriophyes ficus* Cotte and other mites of the fig tree (*Ficus carica* L.) *Bull. Calif. Dept. Agr.*, 28, 266, 1939.

119. Seth, M.L. Transmission of pigeon pea sterility by an eriophyid mite. *Ind. Phytopathol.*, 15, 225, 1962.

120. Seth, M.L. Further observations and studies on pigeon pea sterility. *Ind. Phytopathol.*, 18, 317, 1965.

121. Singh, K., B.S. Dahiya and J.S. Chohan. Evaluation of arhor (*Cajanus cajan*) germplasm lines against the sterility disease in the Punjab. *J. Res. Punjab Agr. Univ.*, 12, 327, 1975.

122. Conners, I.L. *Twentieth Ann. Rep. Can. Plant Dis. Surv. 1940*, p. 104, 1941.

123. Thomas, H.E. and C.E. Scott. Rosette of rose. *Phytopathology*, 43, 218, 1953.

124. Crowe, F.J. 1983. Witches' broom of rose: a new outbreak in several states. *Plant Dis.*, 67, 544, 1983.

125. Doudrick, R.L. Etiological studies of rose rosette. MSc thesis, 101 pp., 1984

126. Amrine, J.W., Jr. and D.F. Hindal. Rose rosette: a fatal disease of multiflora rose. West Virg. Univ. Agr. Forest. Expt. Sta. Circ. 147, 4 pp., 1988.

127. Di, R., J.H. Hill and A.H. Epstein. Double-stranded RNA associated with the rose rosette disease of multiflora rose. *Plant Dis.*, 74, 56, 1990.

128. Amrine, J.W., Jr., D.F. Hindal, T.A. Stasny, R.L. Williams and C.C. Coffman. Transmission of the rose rosette disease agent to *Rosa multiflora* by *Phylocoptes fructiphilus* (Acari: Eriophydae). *Entomol. News.*, 99, 239, 1988.

129. Allington, W.B., R. Staples and G. Viehmeyer. Transmission of rose rosette virus by the eriophyid mite, *Phylocoptes fructiphilus*. *J. Econ. Entomol.*, 61, 1137, 1968.

130. Amos, J. and R.G. Hatton. Reversion of black currants. I. Symptoms and diagnosis of the disease. *J. Pomol. Hort. Sci.*, 6, 167, 1927.

131. Amos, J., R.G. Hatton, R.C. Knight and A.M. Massee. Experiments in the transmission of reversion in black currants. *Ann. Rep. East Malling Res. Sta. Kent*, Suppl. II, 126, 1927.

132. Massee, A.M. Transmission of reversion of black currants. *Ann. Rep. East Malling Res. Sta., Kent, 1951*, 102, 1952.

133. Thresh, J.M. Association between black currant reversion virus and its gall mite vector (*Phytoptus ribis* Nal.). *Nature*, 202, 1085, 1964.

134. Thresh, J.M. Warm water treatments to eliminate the gall mite *Phytopus ribis* Nal. from black currant cuttings. *Jubilee Ann. Rept. East Malling Res. Sta. Kent (1963)*, 131, 1964.

135. Smith, B.D. Experiments in the transfer of the black currant gall mite (*Phytoptus ribis* Nal.) and of currant reversion. *Ann. Rep. Agr. Hort. Res. Sta. Long Ashton, Bristol (1961)*, 170, 1963.

136. Silvere, A.P. Mycoplasma-like organism in association with black currant reversion. *Xth Internat. Congr. Microbiol.*, Mexico, p. 22, 1970.

137. Protsenko, A.E., V.L. Fedotina and N.A. Sergucheva. Mycoplasma-like organisms in the tissue of yellows infected plants. *Prob. Onkol. Teratol. Rast. Leningrad, USSR, Nauka*, 62, 1975.

138. Smith, K.M. *Virus-Insect Relationships*, New York: Longman, Chap. 17.

139. Smith, B.D. A field study of the spread of the black currant gall mite (*Phytoptus ribis* Nal.) and of the virus disease reversion. *Ann. Rept. Agr. Hort. Res. Sta. Long Ashton, Bristol (1962)*, 124, 1963.

140. Thresh, J.M. Black currant reversion disease. *Jubilee Ann. Rept. East Malling Res. Sta. Kent (1963)*, 184, 1964.

141. Pecho, L. The influence of black currant reversion on the content of endogenous gibberellins and inhibitors. *Zahradnictivi*, 17, 105, 1990.

142. Jacob, H. Investigations on symptomatology, transmission, etiology, and host specificity of black currant reversion virus. *Acta Hort.*, 66, 99, 1976.

143. Jacob, H. Undersuchungen zur Ubertragung des virosen Atavismus der Schwarzen Johannisbeere (*Ribes nigrum* L.) durch die Gallmilbe *Cecidophyopsis ribis* Westw. *Z. Pfl. Krankh. Pfl. Schutz.*, 83, 448, 1976.

144. Hutchins, L.M. Peach mosaic—a new virus disease. *Science*, 76, 123, 1932.

145. Hutchins, L.M., E.W. Bodine, L.C. Cochran and G.L. Stout. Peach mosaic. In: *Virus Diseases and Other Disorders with Viruslike Symptoms of Stone Fruits in North America*. US Dept. Agr. Handbook 10, p. 25, 1951.

146. Pine, T.S. Peach mosaic. In: *Virus Diseases and Noninfectious Disorders of Stone Fruits in North America*. US Dept. Agr. Handbook 437, p. 61, 1976.

147. Nemeth, M. *Virus, mycoplasma and rickettsia diseases of fruit trees.* Boston: Martinus Nijhoff Publishers, 841 pp., 1986.
148. Desvignes, J.C. Different symptoms of the peach latent mosaic. *Acta Phytopathol. Acad. Sci. Hung.*, 15, 183, 1980.
149. Desvignes, J.C. Peach latent mosaic and its relation to peach mosaic and peach yellow mosaic virus diseases. *Acta. Hort.*, 193, 51, 1986.
150. List, G.M., N. Landblom and M.A. Sisson. A study of records from the Colorado peach mosaic suppression program. Colorado Agr. Exp. Sta. Tech. Bull. 59, 28 pp., 1956.
151. Wilson, N.S., L.S. Jones and L.C. Cochran. An eriophyid mite vector of the peach mosaic virus. *Plant Dis. Rep.*, 39, 889, 1955.
152. Keifer, H.H. and N.S. Wilson. A new species of eriophyid mite responsible for the vection of peach mosaic virus. *Bull. Calif. Dept. Agr.*, 44, 145, 1955.
153. Jones, L.S., N.S. Wilson, W. Burr and M.M. Barnes. Restriction of peach mosaic virus spread through control of the vector mite *Eriophyes insidiosus*. *J. Econ. Entomol.*, 63, 1551, 1970.
154. McLarty, H.R. Cherry mottle leaf. *Northwest Assoc. Hort. Entomol. Plant Pathol.*, 1, 5, 1935.
155. Reeves, E.L. Mottle leaf of cherries. *Proc. Wash. St. Hort. Assoc.*, 31, 85, 1935.
156. Cheny, P.W. and C.L. Parish. Cherry mottle leaf. In: *Virus Diseases and Noninfectious Disorders of Stone Fruits in North America.* US Dept. agr. Handbook 437, p. 216, 1976.
157. Wilson, N.S. and G.N. Oldfield. New species of eriophyid mites from western North America, with a discussion of eriophyid mites on Populus. *Ann. Entomol. Soc. Am.*, 59, 585, 1966.
158. Borodina, E.E., S.I. Sukhareva, V.A. Shtein-Margolina, L.P. Evgrafova and A.V. Krylov. Biological properties of the pathogen of spotting, a new virus-like disease of cereals. *Mikrobiol. Z.*, 44, 38, 1982.
159. Shukla, D.D., M.J. Frenkel and C.W. Ward. Structure and function of the potyvirus genome with special reference to the coat protein coding region. *Can. J. Plant Pathol.* 13, 178, 1991.

Chapter 9

'VECTORLESS' PATHOGENS WITH UNDISCOVERED VECTORS

Alexander H. Purcell

Alexander H. Purcell is Professor of Entomology at the University of California, Berkeley. He graduated from the United States Air Force Academy in 1964 and received his PhD in Entomology from the University of California, Davis in 1974. He has been on the faculty of the Department of Entomological Sciences at Berkeley since then. Member: AAAS, Entomological Society of America, American Phytopathological Society. Research interests: prokaryotic parasites of the Homoptera, especially plant pathogens, epidemiology of vector-borne plant disease. Address: Dept. of Entomological Sciences, University of California, Berkeley, California 94720, USA.

CONTENTS

ABSTRACT

Numerous virus and virus-like (systemic) pathogens of plants are spread by human propagation of plant materials but otherwise have no known vectors. Where natural spread occurs outside of human dissemination, natural vectors are usually assumed. Little cherry disease and grapevine corky bark and leafroll diseases are 'virus-like' agents that appear to be spread chiefly through human propagation in most parts of the world but where recently natural vectors or other means of viral spread have been discovered in restricted geographic areas. These cases suggest that other 'vectorless' viruses and virus-like agents may similarly have as yet unknown vectors in different plants or in different geographic areas. Thus, instead of guarding against the introduction of viruses, quarantine programs should be aimed at vectors. International cooperation would encourage and speed the research needed to determine essential features of the epidemiology of these agents. Such findings would be very helpful in planning measures to prevent the introduction of efficient vectors into geographic regions where the pathogens are already present but natural spread is not yet a threat.

INTRODUCTION

The transmission by grafting or via interconnections of dodder (vascular parasitic plants of the genus *Cuscuta*) of numerous plant diseases of uncertain etiology suggests that such diseases are caused by viruses or by systemic virus-like pathogens such as mycoplasma-like organisms. This situation seems to occur frequently in plants propagated vegetatively from cuttings or by grafting. For some of these diseases, vectors are known, for many others the mode of spread has not been established.

Where natural spread of such diseases can be observed, a mobile vector is usually assumed to be responsible, and the extent of spread and seriousness of the diseases will determine the practical motivation to search for vectors. But some vegetatively propagated plants have virus or virus-like pathogens that do not appear to spread except via propagation from infected planting material, that is, no mobile vectors are known. Where virus diseases can be satisfactorily controlled by programs of sanitation and certification of propagation materials, there is not much motivation to identify vectors. Indeed, it seems unlikely that we would be able to discover vectors of a disease that does not move from plant to plant!

The detection of natural spread of a virus or virus-like disease thus provides not only the opportunity but also the motivation to discover how the diseases are disseminated. But there also are practical reasons for such studies even for regions where natural spread occurs only by propagation. The primary incentive is to understand the potential threat of inadvertent introductions of vectors to geographic areas where the pathogen occurs without natural spread. The planning of exclusion or quarantine programs against the introduction of exotic plant virus diseases cannot be fully effective until vectors are identified.

Virus diseases of deciduous fruit trees and vines provide numerous examples for which not only vectors but also natural spread are unknown. In using the term 'natural spread', I exclude methods such as grafting or rooting cuttings. Table 9.1 lists such diseases of stone fruits (*Prunus* spp.) and grapevines (*Vitis* spp.) included in the last comprehensive review of virus diseases for these crops in North America [1, 2]. Many other plant diseases could be added to this list by including more recent reports and expanding the geographic area considered. Other vegetatively propagated herbaceous crops such as potato and sugarcane would expand the list even further.

Table 9.1. Virus or virus-like plant diseases of stone fruits and grapevines in North America without documented natural spread [1, 2]

Prunus diseases	*Vitis* diseases
peach calico	rupestris stem pitting
peach red suture	fleck
peach stubby twig	yellow speckle
plum line pattern	leafroll*
cherry pink fruit	corky bark*
cherry black canker	
cherry freckle fruit	

*Leafroll is reported to spread extremely slowly in some localities; corky bark spreads very rapidly in the state of Aquascalientes, Mexico.

In grapevines, leafroll and corky bark diseases occur throughout the world's vineyards. These disease agents are symptomless in rootstock varieties of the species *V. rupestris*, which probably explains their widespread and sometimes common occurrence in vineyards. Consistent spread could not be reliably documented except for that which was attributable to the use of contaminated rootstocks or scion wood.

This picture now has changed both for the grape diseases and for little cherry disease. Recent discoveries of vectors of the agent of little cherry disease (LCA) of sweet cherry (*Prunus avium*) and of closteroviruses that may be the pathogens of grapevine leafroll and corky bark diseases illustrate how the vectors of even very widespread viruses can go undetected because active natural spread was not noticed or was restricted to small geographic regions.

LITTLE CHERRY DISEASE

For many years, the rapid spread of little cherry disease in western Canada was unexplained [3]. Searches for vectors emphasized arthropods, chiefly those taxa already incriminated as vectors of plant viruses [4]. Disease indexing of ornamental flowering cherries revealed that the causal agent of this disease was present in most accessions of flowering cherry (*P. serrulata*) [2, 5]. Flowering cherries were symptomless hosts of LCA, or at least of a pathogen that induced identical symptoms in indicator varieties of sweet cherry [5] (Plate 9.1). The lack of spread of little cherry disease throughout the rest of North America, except for its occasional appearance in the nearby state of Washington, was attributed to the restriction of efficient vectors to western Canada. There were reports of some natural spread in England [6].

The recent discovery of a mealybug vector of LCA, the apple mealybug (*Phenacoccus aceris*), seems to explain satisfactorily the pattern and rate of spread of little cherry [7]. The widespread occurrence of LCA in ornamental flowering cherries around the world without the appearance of new infections in sweet cherry would be explained by the lack of natural vectors in most parts of the world, compounded by the latent (symptomless) nature of the agent in flowering cherry. This finding also suggests efforts that should be made to prevent further spread within Canada, namely that insecticides be applied to prevent vectors moving from freshly removed infected plants. Moreover, the mealybug vector must be prevented from expanding to adjacent regions in order to stop the introduction of natural spread of little cherry to the adjacent western states of the United States.

Like any initial discovery, this one raises further questions. First of all, can mealybug transmission be confirmed by others? Previous reports of leafhopper transmission of LCA [8] are in conflict with the observed patterns of spread, and the many years without confirmation that have elapsed since this report challenge its acceptance. What other mealybugs are capable of transmitting LCA? Are strains of LCA in flowering cherry also transmissible to sweet cherry? The answers to these questions and others will be needed to insure rational approaches to preventing the spread of little cherry disease to other regions.

CORKY BARK AND LEAFROLL DISEASES OF GRAPE

Corky bark (Plate 9.2) or diseases very similar in appearance and cultivar susceptibility have been noted in all major grape-growing regions of the world. In California, the random occurrence of corky bark within vineyards and the findings that many cultivars were symptomless suggested that most disease spread was due to propagation from infected rootstocks or scion wood [9].

Leafroll disease of grapevines (Plate 9.3) is far more widespread than corky bark and also was assumed to be a virus disease for which spread could be prevented entirely by the use of disease-free planting stocks. Heat therapy provided a way to eliminate these pathogens from parental stocks [2]. Its widespread occurrence was attributed to the long incubation period (18 months) in most varieties, the mildness of symptoms for most lines of the pathogen or for some varieties, and ease of graft transmission. No vectors were evident [2], and natural spread was reported to be slight [10] or none [11], depending upon the grape-growing region.

Over the past decade, there have been numerous reports (see *Phytopathologia Mediterranea* Vol. 24, 1985) of various virus particles from leafroll-diseased grapevines. The initial association of closterovirus-like particles with leafroll disease was reported in 1979 [12]. It appears that the leafroll syndrome may be caused by a complex of viruses and virus strains.

The etiological role in leafroll disease of the transmitted viruses remains unresolved [10, 13-14]. The transmission of at least some of these putative closteroviruses by mealybugs (Homoptera: Pseudococcidae) implicates this group of insects as possible vectors, although definitive proof is still lacking. A closterovirus named grapevine virus A was transmitted by the long-tailed mealybug, *Pseudococcus longispinus*, from grape to grape and to herbaceous plants [14]. In South Africa, leafroll disease recently has been observed to spread naturally from grapevine to grapevine, and transmission of closterovirus by the mealybug *Planococcus ficus* was reported [13]. Transmission of the corky bark disease from grape to grape by *P. ficus* was recently reported from Israel [15].

Corky bark is not known to spread rapidly, except in central Mexico. In the brandy-producing region of the small state of Aquascalientes, Mexico, grapevines have very high rates of infection with corky bark [1, 16]. The symptomless infection of native American grape species with corky bark, along with the very high rates of spread [16], suggest not only that an efficient vector is common in the state of Aquascalientes, Mexico, but that this area may be the center of origin of the disease. The identification of the vectors responsible for the rapid spread of the disease in Mexico would aid appraisals of the risks that such vectors might pose to viticulture in the United States or elsewhere and what precautions might be taken to prevent their introduction. At present, most quarantine efforts in California, along the nothern border of

Mexico, are directed at preventing the entrance of infected plants. But in the case of corky bark disease, preventing the spread of the vector would be more logical because the corky bark agent is already present in California. On the other hand, efficiently transmissible strains of corky bark agent may largely explain the high rates of spread in central Mexico. The main point is that unless these factors are better understood, not only California but other parts of the world may be at risk from the active natural spread of corky bark. This could drastically increase the importance of a disease that is now relatively minor because clean planting stock programs can control the disease effectively in the absence of natural spread [17].

A CASE FOR INCREASED INTERNATIONAL INVOLVEMENT

There may be reasons other than the absence of efficient vectors to explain why virus diseases do not spread in areas with diseased planting stocks. Strains of virus may lose their vector transmissibility following prolonged vegetative or mechanical transmission. This has been noted for numerous potyviruses. Vegetatively propagated crops may acquire rare infections of viruses or virus-like agents that are then maintained by propagation but cannot subsequently serve as adequate sources of inoculum for vector transmission. These or other explanations are merely speculation until more facts are available. But until more research is done, such facts will never be available.

As I pointed out previously, countries or regions that experience significant levels of natural spread of virus diseases have the practical motivation to undertake pathological and epidemiological studies of these problems. Unfortunately, some countries may not have the resources necessary to sustain the needed research. The crops involved may not be important enough in the countries in which natural spread occurs to justify expenditures towards the needed research. In contrast, countries that lack natural spread of the disease also do not have a compelling motivation or urgency in solving 'someone else's problem'. What this latter viewpoint overlooks is that a minor problem in one country may become a large problem in another when transplanted and that ultimately the cheapest solution is prevention.

The appearance of sharka virus of stone fruits in western Europe raised instant concern and major efforts at eradication and containment. It also heightened awareness and caused increased quarantine measures to prevent its introduction to unaffected but ecologically isolated regions. This reaction was based largely on knowledge of how the disease spreads, its host range, and other aspects of its epidemiology. There are many other viruses that may be equally threatening but our knowledge of them does not permit a quick response or, perhaps more importantly, does not encourage logical preventative measures. It is at the international level that we must shape our perspectives of diseases that may already be globally widespread because of plant propagation but are only actively spreading in a few or single regions. Individual scientists usually lack the perspective of internationally oriented organizations. Funding of plant protection research logically is directed towards problems that are already known to be serious rather than investigating diseases that typically are regarded as academic curiosities that may or may not become more serious problems. Furthermore, it is obvious that research on the transmission of a plant disease agent should not be made in localities where the vectors of the agent do not occur. To overcome these difficulties, we must increase international collaboration.

REFERENCES

1. Pearson, R.C., and A.C. Goheen. (eds.). *Compendium of Grape Diseases*. St. Paul, Minnesota: APS Press, 93 pp., 1988.
2. Pine, T.S., R.M. Gilmer, J.D. Moore, G. Nyland and M.F. Welsh. (eds.). *Virus Diseases and Noninfectious Disorders of Stone Fruits in North America*, US Dept. Agric. Handbook 437, Washington, DC, 433 pp., 1976.
3. Welsh, M.W., and P.W. Cheney. Little cherry. In: *Virus Diseases and Noninfectious Disorders of Stone Fruits in North America*, US Dept. Agric. Handbook 437, Washington, D.C., pp. 231–237, 1976.
4. Forbes, A.R., C.K. Chan, J. Raine and R.D. McMullen. Aphids trapped in Okanagan cherry orchards and the failure of nine species to transmit little cherry disease. *J. Entomol. Soc. Brit. Columbia.*, 82, 39, 1985.
5. Reeves, E.L., P.W. Cheney and J.A. Milbrath. Normal-appearing Kwanzan and Shirofugen oriental flowering cherries found to carry a virus of the little cherry type. *Plant Dis. Rep.*, 39, 725, 1955.
6. Posnette, A.F. Virus diseases of cherry in England. I. Survey of diseases present. *J. Hort. Sci.*, 29, 44, 1954.
7. Raine, J., R.D. McMullen and A.R. Forbes. Transmission of the agent causing little cherry disease by the apple mealybug *Phenacoccus aceris* and the dodder *Cuscuta lupiliformis*. *Can. J. Plant Pathol.*, 8, 6, 1986.
8. Wilde, W.H.A. Insect transmission of the virus causing little cherry disease. *Can. J. Plant Sci.*, 40, 707, 1960.
9. Beukman, E.F., and A.C. Goheen, Grape corky bark. In: *Virus Diseases of Small Fruits and Grapevines*. W.B. Hewitt and N.W. Frazier, eds. Davis, California: Univ. of Calif. Div. of Agric. Sci., pp. 209–212, 1970.
10. Tanne, E. New data on grapevine leafroll disease and its agent. *Phytopathol. Mediter.*, 24, 88, 1985.
11. Goheen, A.C. Grape leafroll. In: *Virus Diseases of Small Fruits and Grapevines*. W.B. Hewitt and N.W. Frazier, eds. Davis, California: Univ. of Calif. Div. of Agric. Sci., pp. 207–209, 1970.
12. Namba, S., Y. Yamashita, Y. Doi, K. Yora, Y. Terai and R. Yano. Grapevine leafroll virus, a possible member of closteroviruses. *Ann. Phytopathol. Soc. Jpn.*, 45, 497, 1979.

13. Englebrecht, D.J., and G.G.F. Kadorf. Association of a closterovirus with grapevines indexing positive for grapevine leafroll disease and evidence for its natural spread in grapevine. *Phytopathol. Mediter.*, 24, 101, 1985.

14. Rosciglione, B., M.A. Castellano, G.P. Martelli, V. Savino and G. Cannizzaro. Mealybug transmission of grapevine virus A. *Vitis*, 22, 331, 1983.

15. Tanne, E., Y. Ben-Dov and B. Raccah. Transmission of the corky bark disease by the mealybug *Planococcus ficus*. *Phytoparasitica*, 17, 55, 1989.

16. Teliz, D., P. Valle, A.L. Goheen and S. Luevano. *Proc. 7th Meeting ICVG*, Niagara Falls, Canada, pp. 51–64, 1980.

17. Hewitt, W.B. From virus-like to virus diseases of grapevines: Some unresolved problems including immunity and ideas for researching them. *Pytopathol. Mediter.*, 24, 1, 1985.

Chapter 10

NON-INFECTIOUS DISEASES OF UNCERTAIN ETIOLOGY

Karl Maramorosch

CONTENTS

ABSTRACT

Two diseases of coconut palms are described: bristle top of Jamaica and Guam, and lethal decline of Mauritius. No pathogens could be detected in these diseases. Inadvertent effects of herbicides or plant hormones used in areas adjoining coconut palms were incriminated as likely causes of the observed disease symptoms.

INTRODUCTION

Whenever plant pathologists observe a disease and are unable to find a causative agent, the disease qualifies as a disease of unknown or uncertain etiology. In certain instances causative agents have been described but not confirmed by careful follow-up studies so that the diseases had to be reclassified as uncertain etiology diseases. Kerala coconut wilt (synonyms: root rot, wilt disease of Cochin-Travancore) is a good example. This disease has been known in South India since 1876. In 1948 investigations of its etiology became part of the program of the Central Coconut Research Station at Kayangulam, under the administration of the Indian Central Coconut Committee. The disease is present in a large area under coconut production in Kerala and the reduction in copra production in affected areas is quite serious. Its spread into Tamil Nadu State as well as into new areas in Kerala seems to indicate that it is an infectious disease. In the 1960s it was reported that the disease is caused by a mechanically transmissible virus. More recently claims have been published that the disease is caused by mycoplasma-like agents and two insects have been incriminated as vectors. Transmission of the MLOs to periwinkle through *Cassythia filiformis* has also been claimed [1]. However, none of these assumptions could be confirmed and the disease remains thus unfortunately a disease of uncertain etiology [2].

Differences between uncertain etiology diseases are established on the basis of characteristic symptoms, forecast of the cause of the disease, and pathogenesis, i.e., the development of the disease from appearance of first symptoms to conclusion. Pathogens such as fungi, bacteria, viruses, viroids, mycoplasma-like agents, nematodes, and flagellates account for most known infectious plant diseases. Lack or abundance of macro- and microelements in the soil, air pollution, growth hormones, and weed killers cause physiologic, non-infectious plant diseases. Two presumptive instances of the latter, as yet uncertain etiology diseases are described in this chapter.

BRISTLE TOP DISEASE OF COCONUT PALMS

Bristle top disease (synonyms: distorted nut, Chinese nuts, fused leaf condition, coconut palms with fused pinnae) has been observed in Jamaica since 1960. During a world-wide survey of coconut diseases, conducted in 1963 under the auspices of the Food and Agriculture Organization of the United Nations, I noticed fused pinnae and distorted nuts in several coconut palm plantations on the island of Jamaica [3]. Leaves of affected palms assumed the form of juvenile leaves although they were produced in the crown of a bearing tree. Pinnae were fused (Plate 10.1) and they remained in this condition until they died. No spots of any kind have been seen on such juvenile leaves. The bristle top appearance of palms was due to these distorted leaves. Inflorescences were formed profusely and more buttons were set and retained than on normal female inflorescences. The most striking feature of the disease was the severe distortion of nuts (Plate 10.2). These were of various sizes, devoid of meat but almost completely filled with the husk. A blackening of the inside was often seen upon splicing of the husk.

Some uncertainty and confusion existed at the time of the 1963 FAO survey as to the cause of bristle top disease, because the condition was seen not only in areas where herbicides had been applied, but also in plantations where there was no known record of herbicide use. However, a delayed effect of approximately eight months was required for the development of symptoms, following the application of weed killers. Droplets of sprayed herbicides could be carried by wind to areas where no deliberate herbicide applications had been made. Since a similar symptomatology, with fused pinnae and distorted nuts, was linked to aerial applications of herbicides on the Pacific island of Guam [3], bristle top disease is most likely a non-infectious physiological disorder, caused by compounds such as 2,4D, 2, 4, 5-T, and monuron (Telvar).

LETHAL DECLINE OF COCONUT PALMS ON MAURITIUS

In recent years a coconut palm disease has been observed on the island of Mauritius in the southern Indian Ocean. No coconut plantations existed on Mauritius in 1985 when I conducted a survey there at the request of the Mauritius government and FAO. Palms were scattered in gardens, on beaches, near hotels, and along roads in the coastal areas. The reason for the requested survey was the assumption that the decline might be caused by lethal yellowing disease or a rapidly spreading infection and that it would completely wipe out the remaining trees. The island had to import coconuts from other areas, including some smaller islands as well as Sri Lanka, to satisfy local demand for the nuts.

Plans were also made to expand considerably the planting of coconut palms on Mauritius. Since some of the disease symptoms resembled those of lethal yellowing it was very important to establish whether the devastating MLO disease has, in fact, reached the island far removed from the known lethal yellowing areas in West Africa, the Caribbean, and Mexico.

The symptoms observed by me in March 1985 [4] consisted of necrosis of inflorescences still ensheathed in their spathes (Plate 10.3); male flowers turning brown at the tips of the rachillae and easily breaking off (Plate 10.4); dwarfed nuts, occasionally found dropping off; drooping and drying of fronds. Only occasionally the bright yellow discoloration, typical of lethal yellowing, was seen.

Although several similarities were noticed between the symptoms described above and those of lethal yellowing, no MLOs were detected when samples from diseased palms were prepared from inflorescences and examined by electron microscopy techniques [5]. Nucleic acids were analyzed by two-dimensional electrophoresis, eluted from three positions, and samples treated with RNase T1, polynucleotide kinase, and gamma[32p] ATP and then fingerprinted. No evidence was found for circular RNAs associated with viroids, nor for small double-stranded nucleic acids that might be associated with infectious agents. The absence of microbial pathogens and viroid-type nucleic acids in freeze-dried leaf material pointed to a non-infectious etiology of the disease. This hypothesis was further strengthened by the observation on the island of coconut palms at Pamplemousses, along the outer rim of the Botanic Garden, where 100% of the coconut palms were affected and all were in the same stage of the disease—as if deliberately infected on a single day. A few feet from the row of palms there started a large field under sugarcane. In practically all other locations on Mauritius where the disease occurred, the palms grew in close proximity to sugarcane as well. This observation provided a clue for the presumptive cause of the disease. Weed killers as well as chemicals used to control the flowering of sugarcane, combined with prolonged droughts, might have been responsible for the necrosis of the inflorescences of coconut palms especially prevalent on dwarf varieties. The disease outbreak was preceded by a prolonged, severe drought, which influenced physiological conditions, perhaps lowering the resistance of the trees to certain growth substances and weed killers used on the adjoining sugarcane plantations. The similarity of disease symptoms on the island of Mauritius, Guam, and Jamaica, where the coconut palm disorders were earlier linked with applications of weed killers, favors the assumption:

CONCLUSIONS

Although the coconut palm disease on Mauritius is, apparently, non-infectious and certainly not caused by lethal yellowing disease MLOs, constant vigilance is needed to intercept the possible introduction of lethal yellowing disease to Asia and the Pacific area. Strict quarantine regulations ought to be enforced to prevent the importation of coconut seed nuts from the Western hemisphere and from West Africa. The rapid spread of lethal yellowing disease in Mexico during the 1985–1989 period [6] illustrates how this disease can spread over a large area in a short time, most likely assisted by winds that favor the movement of the vector, and by human activity. In parts of the Yucatan, losses of 100% are now anticipated. The proper diagnosis of the causes of plant diseases of cultivated as well as of non-cultivated plants and weeds is the first step to provide proper control measures and prevent serious losses to food and fiber crops.

REFERENCES

1. Solomon, J.J., M.P. Govindankutty, K. Mathen, C.P.R. Nair, M. Sasikala and N.C. Pillai. Evidences towards a mycoplasmal etiology for the root (wilt) disease of coconut in India. Abstracts, Indo-US Workshop on Viroids and Diseases of Uncertain Etiology, IARI, New Delhi, pp. 20–21, 1989.
2. Summanwar, A.S. Coconut root (wilt)—disease of uncertain etiology. Abstracts, Indo-US Workshop on Viroids and Diseases of Uncertain Etiology, IARI, New Delhi, p. 19, 1989.
3. Maramorosch, K. A survey of coconut diseases of unknown etiology. Rome: FAO, 38 pp + 61 color figs., 1964.
4. Maramorosch, K. Report to FAO/UN Rome on Consultancy on Coconut Diseases in Mauritius. FAO-AGP 57468-TCP-4403 A, 1985.
5. Maramorosch, K. Lethal decline of coconut palms in Mauritius: an enigma. In: *Vistas in Plant Pathology*. A. Varma and J.P. Verma, eds. New Delhi: Malhotra Publ. House, pp. 185–190, 1986.
6. Robert, M.L., V.M. Loyola-Vargas and D.V. Zizumbo. Lethal yellowing in Mexico. *Bull. BuroTrop*, 1, 13–14, 1991.

Chapter 11

BREEDING COCONUTS RESISTANT TO MLO, VIROID AND VIRUS DISEASES

Hugh C. Harries

Hugh Harries is a consultant specializing in coconut, date and oil palms. After graduation (BSc, London University, 1962) he worked as a plant breeder in Jamaica, 1967–1978; Thailand 1978–1983; Papua New Guinea, 1983–1988; and Tanzania, 1990 to date. Currently writing the third edition of Coconuts for Longman's Tropical Agriculture series. Registrar, International Society Hort. Sci. Registration Authority for Coconut Cultivars. Member: Gen. Society, Association of Applied Biologists, Inc. Society of Planters, International Palm Society. Address: 17 Alexandra Road, Lodmoor Hill, Weymouth, Dorset DT47QQ, England.

CONTENTS

THE DISEASES

MLO Diseases

Epidemic coconut diseases associated with mycoplasma-like organisms (MLOs) are known as lethal yellowing in parts of the Caribbean [1, 2] and Central America [3], Kaincopé disease [4], Cape St. Paul wilt [5, 6] and Kribi disease [7] in West Africa, and lethal disease in East Africa [8]. Root wilt in India [9] has also been associated with MLO and there are other diseases such as Awka [10] disease in Nigeria and 'unknown disease' (Doença desconhecida) [11] in Mozambique which may prove to be the same. (Amarilliamento Létal in Mexico is merely a Spanish translation of the English name used in the Caribbean and Florida and is not a different disease.) Another, less conclusively similar disease may be Natuna wilt [12] in Indonesia. Although epidemic MLO diseases have not been confirmed from southeast Asia, electron micrographs have shown MLO in coconut palm in both Indonesia and Malaysia [13]. This tends to support the argument that this type of disease originated in that part of the world [14] where coconuts were domesticated [15], giving rise to resistant varieties such as the Malayan Dwarf.

Viroid Diseases

There are two serious diseases with a viroid etiology, both of which occur in the northwest Pacific [16, 17]. Tinangaja disease of coconut palm in Guam is considered to be identical to Cadang-cadang in the Philippines and caused by the same pathogen. At one time it was thought that these diseases did not extend over such a wide geographical area as do those associated with MLO. Now, continued survey work appears to locate viroids elsewhere in the Pacific region from locations where no epidemic coconut disease has been reported, for example, in oil palm and coconut on the Solomon Islands [18] and in coconut on Samoa [19]. It remains to be seen what will be found when surveys are extended to Africa and America, whether viroids and MLO can coexist in any given coconut population or even within a single palm.

Virus Diseases

Not unexpectedly, perhaps, lethal yellowing, Cadang-cadang, and root wilt had been considered virus diseases, before MLOs and viroids had been recognized. In none of these cases has any causal relationship been (satisfactorily) demonstrated. In contrast, the suspected virus-like nature of foliar decay in the southwest Pacific [20] has recently been confirmed following artificial inoculation [21]. It is of particular interest that the putative insect vector of foliar decay in Vanuatu [22] is of the same genus as that of lethal yellowing in Florida [23]. Yet in all other respects the two diseases appear to be distinct. Earlier reports of successful mechanical inoculation techniques with lethal yellowing disease [24] might need to be re-assessed, to take advantage of new knowledge. Virus-like particles have been reported in association with lethal yellowing [25] as well as root wilt [26] and with Cadang-cadang [27]. The rod-shaped particles resembling viruses which were found in Cadang-cadang infected palms were not thought to be pathogenic because they were also seen in healthy palms and in palms from outside the diseased area. As in the case of viroids, it may be asked whether viruses can be found in the same populations or even the same palms as MLO.

RESISTANT VARIETIES

Coconut varieties are not so uniform that all the palms in any given population are alike. Depending on the characteristics under examination they show greater or lesser variability. Differences in height, flowering pattern, and fruit color between tall and dwarf varieties are commonly recognized. Between tall varieties particularly, there are also distinct differences in fruit components (amount of husk, shell, water, and meat) and germination rates [28, 29]. But individual palms within any variety are not always easily categorized and this applies to disease resistance. Where some varieties show high resistance, others are intermediate and some are low. The individual palm may be highly resistant, highly susceptible, or somewhere in between. Only in the case of lethal yellowing has it been suggested that coconut populations showing wild type characteristics are more susceptible than those showing domestic characteristics [30].

The Malayan Dwarf is perhaps the best-known example of a successful disease-resistant coconut variety and the basis for a successful breeding program against lethal yellowing [31]. It was originally introduced to Jamaica not for disease resistance, but because of its precocity. Early-bearing palms were needed to recover production quickly after a hurricane [32]. At that time the disease was less important and there was no clue that resistance was possible. Yet this resistance has operated in Jamaica for almost 50 years and although a few isolated cases of high losses have been reported the Malayan Dwarf type of resistance is still the best choice for replanting [25].

The reported loss of Malayan Dwarf palms in Ghana is much too low to make any statement about lower resistance or loss of resistance of this variety in West Africa [33]. Recent field trials have failed to confirm this loss, apparently because the disease has bypassed trial areas. Dwarf varieties in Cameroon, which are not the same as Malayan Dwarf, have been shown to be more susceptible. On the basis of symptoms among susceptible palms, but in the absence of results from field exposure trials, it was anticipated that the situation in East Africa would follow the Caribbean example. It has not done so and a possible genetic explanation has been put forward [34]. In effect, no varieties have shown the high resistance of the Malayan Dwarf; local varieties have shown as good resistance as introduced varieties, possibly better. This is similar to the Vanuatu virus situation. In the case of foliar decay in Vanuatu, local varieties are considered to be resistant and no disease was suspected until after foreign varieties had been introduced [35]. There was no suggestion that the disease itself had been introduced and both a local vector and an alternate host have been identified. Hybrids have shown tolerance. For viroid diseases such as Cadang-cadang there is apparently no local or introduced resistant variety. The disease occurs on the Pacific coastlines of northern and eastern islands where wild-type coconuts are found [36] and not in the southern or western coastlines where domestic varieties predominate. The differences in varietal response which have been identified [37] may be directly comparable to differences already observed in lethal yellowing.

BREEDING FOR RESISTANCE

The Philippines is the most important coconut-growing area and Cadang-cadang therefore is the most important disease. In contrast, the coconut-growing countries of America and Africa are of minor commercial importance. This can be directly attributed to the historically late introduction of coconuts to those areas. For the same reason the variability of the local coconut varieties is restricted. It is perhaps for this reason that the most successful disease resistance breeding effort to date is that of the Research Department of the Coconut Industry Board of Jamaica. Between 1959 and 1967 the CIB made worldwide germplasm collections and established comprehensive variety trials to seek resistance to lethal yellowing as good as that shown by the Malayan Dwarf [38]. From 1968 to 1978 the CIB research tested local, introduced, and hybrid material and concentrated on finding ways to produce the resulting hybrids commercially [39]. Since 1978 testing has continued. The ultimate step will be to produce true breeding varieties, either by selection and hybridization or by tissue culture.

Breeding for Lethal Yellowing Resistance

The research into lethal yellowing will be taken as an example since it shows how breeding research must be backed up by research into the nature of the pathogen. From a practical point of view, the detailed knowledge of the disease has not yet provided feasible control measures, and breeding has been the only possible answer. To be successful, a breeding program demands efficient selection methods. In case of disease resistance this generally implies a screening procedure in which a large number of plants can be subjected to a high infection pressure at a developmental stage in which symptom expression can be rapidly and confidently determined. It must be admitted that in breeding coconuts for resistance to lethal yellowing practically none of these criteria can be met. Planting density is low (80–140/ha); symptom expression is delayed (3 to 6 months after infection in young palms and 7 to 15 months in mature palms); very young palms (less than 18 months old) rarely become infected; full symptom expression is seen only in bearing palms; palms cannot be artificially infected; and infection pressure appears to depend on the incidence of disease in the immediate neighborhood of the trial (over which there is no control). Lethal yellowing is perhaps the most virulent of coconut diseases and while the commercial coconut grower may suffer from the ravages of this disease it gives advantage to the breeder in comparison with diseases of slow mortality, such as Cadang-cadang or root wilt.

The Jamaica Tall, the Panama Tall and the Malayan Dwarf have very different levels of susceptibility or resistance to lethal yellowing. These three types of coconut are particularly suitable as standards against which other varieties and F_1 hybrids can be compared. It would appear that some varieties share both desirable and undesirable characteristics with one or more of the standard varieties. None has yet proved superior to the locally selected Malayan Dwarf. Having investigated the local varieties and determined that only one, the Malayan Dwarf in its three color forms, was highly resistant to lethal yellowing, the second stage of the breeding program was to introduce a representative sample of foreign varieties to test alongside these standard varieties in field exposure tests.

Coconut breeding proceeds on the basis that geographical ecotypes are distinct; but there are many exceptions to this general rule. Effectively dwarf and tall ecotypes are crossed to produce dwarf × tall F_1 hybrids. The reciprocal tall × dwarf crosses have been less widely tested. This is because it is not so easy to use the tall type as a seed parent and still ensure the authenticity of the cross. Tall × tall crosses are coming into interest where there is dissatisfaction with dwarf × tall hybrids. It is too soon to say how successful they might be. The dwarf × dwarf crosses are generally disregarded although, historically, the Malayan Dwarf × Fiji Dwarf was first made over 50 years ago. Almost without exception the crosses must be made before there is any real knowledge of what to expect. Hybrid trials not only test the hybrids but also provide essential information about the combining ability of the parents.

Field Testing for Resistance

By replicating test varieties in diseased areas throughout commercial coconut plantations each variety is given the greatest chance to express its reaction to disease. However, this may leave only small numbers of palms for use as pollen parents. It is therefore necessary, if time is not to be lost, for crosses to be made in advance of knowledge of the performance of the parental populations. This applies to yield and other characters as well as to disease. In lethal yellowing, the only member of the resistant group, the Malayan Dwarf, is easily maintained true to type, being 80–90% self-pollinating and largely colour-marked. In the less resistant group of varieties it remains to be seen whether continued losses will put some into the highly susceptible group. Comparison with that group suggests that a real degree of resistance exists. The disease tends to move out of any area after a few years and palms left behind are either resistant or may have escaped becoming infected. To find out which is which, progenies of these palms have to be planted in areas that are, or that become, actively diseased. The degree of loss will then confirm the type of the original parent. This step has been taken in Jamaica, but recent hurricanes will delay results. There has been a reduction in the extent and activity of disease in Jamaica, as a result of the successful replanting program [40]. Although this is good for the farmer it does not help the breeder. This serves to emphasize the need to find artificial tests of resistance.

When the palms that escaped infection have been identified and discarded the factors for resistance in those remaining are likely to be heterozygous, due to cross-pollination. Any attempt to produce a uniformly highly resistant, cross-pollinating variety will be possible only if there is access to large backcross populations and the means by which to screen them. Moreover, since the varieties under consideration are not superior to the Malayan Dwarf in yield characteristics it is unlikely that any variety will be improved simply by selection. A backcross program will again be needed. The production of F_1 hybrids which are becoming highly regarded in their own right is the first step in a backcross program.

In the case of lethal yellowing this was done over 20 years ago and practical F_1 hybrids have been grown commercially for 10 years or more. Possible loss of resistance, though reported, has not yet been confirmed and in no way devalues the effort or the achievement. In the MLO diseases in Africa, screening in West Africa has failed to produce practical results, seemingly because the disease did not attack field trials as expected. The reputed loss of dwarf resistance, which can more readily be attributed to poor growing conditions, has also not been satisfactorily resolved. On the other side of the continent lethal disease has not conformed to expectations based on Jamaican experience. The introduced varieties and F_1 hybrids have appeared to show no better resistance than some local selections.

Hybrid Production

Research and development organizations in many countries see F_1 hybrid planting material and better agronomic practices as the best way to improve coconut yields. Early hybrid seed production methods could not produce enough planting material. Low seed set per inflorescence even under natural conditions means that hybridization using isolation bags is inefficient for any but experimental quantities of seed. Hybrid seed gardens, consisting of rows of seed and pollen parents, have to balance the maximum number of seed from the one with an adequate amount of pollen from the other. It was for this reason that assisted pollination was introduced, whereby additional pollen was brought into an interplanted seed garden. Problems with this system are that tall and dwarf parents have different spacing requirements, the onset of flowering will be different and, once the hybrid seed garden is established, only one type of hybrid can be produced. Therefore, other hybrid seed gardens have to be established if a choice of F_1 hybrid planting material is to be offered to the farmer. Since each seed garden must be isolated, problems of land space and supervision over scattered sites may arise.

The general susceptibility of tall varieties to lethal yellowing makes the establishment and life expectancy of an interplanted seed garden very questionable. Many or most of the pollen parents may die. Instead, use can be made of isolated plantings of dwarf palms, at 240 plants per hectare. These can be planted where they do not already exist. When the dwarf palms are nearly all bearing, in the fourth or fifth year, emasculation is carried out daily. Pollen is collected from the selected surviving tall parents. Many different tall parents may be deliberately planted together at another site; in Jamaica, for example, they are in existing variety trials. Pollen might be imported if processing facilities are available elsewhere. This would make many new F_1 hybrids immediately available. Should a different dwarf × tall hybrid combination be desired, the change can be made in a matter of weeks.

There is a further advantage to this method of mass controlled pollination in isolated seed gardens. It gives the plant breeder the ability to produce large numbers of seedlings which may segregate to give desirable recombinations. For this purpose, the seed garden is not limited to dwarf types. Instead, the plant breeder emasculates fields of hybrids as seed sources to produce backcross, top cross, and later generation test material. Two different F_1 hybrids could be crossed to produce a population combining the qualities of three or four different parental types. Selections from F_3 or F_4 generations could provide pollen for use on their sibs to attempt to fix desirable characters in a true breeding variety. The ability to produce large quantities of breeding material will facilitate genetic studies of inheritance. The only problem will be space to plant and test the progenies for disease resistance, when plant habit is likely to vary due to dwarf parentage. It is here that the plant pathologist must come to the aid of the plant breeder by devising methods to reliably and rapidly screen large numbers of young plants for disease resistance.

CONCLUSION

Breeding work has been aimed at producing dwarf × tall hybrids. These have many technical advantages for seed production as well as important ones in precocity and yield for the farmer. In the case of the Maypan, bred specifically for lethal yellowing resistance, definite success can be claimed, both in Jamaica and in Florida. Other hybrids which were produced for yield rather than disease resistance, such as those produced by IRHO in the Ivory Coast, have not been tested extensively against that disease in the Caribbean. Trials in West Africa appear to be inconclusive, whereas those in East Africa have definitely been unsuccessful. One hybrid has shown some success against foliar decay in Vanuatu but none have yet been found with resistance to Cadang-cadang. It has not been possible to test Maypan outside Jamaica and Florida because of the small, but unacceptable, risk of carrying the disease or its vector on seednuts even when these have been fumigated.

Few of the available disease control methods used on other crops are applicable to coconut diseases. Chemicals, such as antibiotics for MLO disease control, are expensive and are unlikely to eradicate infection or realistically to limit its spread. In the case of perennial crops, one application of a chemical does not given continuous protection. Cultural improvements, such as the use of fertilizer and appropriate cultivations, may moderate one disease, as in the case of root wilt, but possibly encourage lethal yellowing. Annual or seasonal crop rotation is hardly practial in a perennial tree crop. A method must therefore be found which is cheap and economical, can be applied as a routine procedure, is safe and non-toxic, and, most vital, gives protection throughout the life of the plant. Resistant varieties come closest to meeting these requirements. Seed is comparatively cheap, and the cost of breeding and maintaining a variety is low when offset against its productive life. Protection operates every time a seed is sown and lasts indefinitely. Furthermore, biological control of disease introduces no hazards to health. But as pathogens are also genetic organisms they are inherently variable. A new race can arise, and resistant varieties are no panacea. However, if a combination of control methods is used, that is to say, resistant varieties, vector control, eradication of alternate hosts, and general plantation hygiene, one will complement another and successfully reduce the risk. A farmer can then spend time and money producing a good crop rather than being nursemaid to a sick one.

REFERENCES

1. Howard, F.W. and C.I. Barrant. Questions and answers about lethal yellowing disease. *Principes*, 33(4), 163–171, 1989.
2. Plavsic-Banjac, B., P. Hunt and K. Maramorosch. Mycoplasma-like bodies associated with lethal yellowing disease of coconut palms. *Phytopathology*, 62, 298–299, 1972.
3. McCoy, R.E. What's killing the palm trees? *National Geographic*, 174, 120–130, 1988.
4. Gianotti, J., et al. Mise en culture de mycoplasmes a partir de racines et d'inflorescences de cocotiers atteints par la maladie de Kaincope. *Oleagineux*, 30, 13–18, 1975.
5. Dabek, A.J., C.G. Johnson and H.C. Harries. Mycoplasma-like organisms associated with coconut palms in West Africa. *PANS* 22(3), 354–358, 1976.
6. Dabek, A.J. Electron microscopy of Kaincopé and Cape St. Paul wilt diseased coconut tissue from West Africa. *Phytopathol. Z.*, 88, 341–346, 1977.
7. Dollet, M., J. Gianotti, J.L. Renard and S.K. Ghosh. Etude d'un jaunissement letal des cocotiers au Cameroun: la maladie de Kribi. Observations d'organismes de type mycoplasmes. *Oleagineux*, 32, 317–322, 1977.
8. Schuiling, M., F. Nienhaus and D.A. Kaiza. The syndrome in coconut palms affected by a lethal disease in Tanzania. *Zeitschrift fur Pflanzenkrankheiten und Pflanzenschutz*, 88, 665–677, 1981.
9. Solomon, J.J., M.P. Govindankutty and F. Nienhaus. Association of mycoplasma-like organisms with the coconut root (wilt) disease in India. *Zeitschrift fur Pflanzenkrankheiten und Pflanzenschutz*, 90, 295–297, 1983.
10. Ekpo E.N. and E.E. Ojomo. The spread of lethal coconut diseases in West Africa. *Principes*, 34, 143–146, 1990.
11. Quadros, A.S. A 'doença desconhecida' do coqueiro na Zambezia. *Revista Agricola Mocambique*, 14, 33–34, 1972.
12. Hunt, P. A coconut disease of uncertain etiology in Indonesia. 4th Meeting International Council on Lethal Yellowing, Florida, 1979.
13. Turner, P.D., P. Jones and R.M. Kenten. Coconut stem necrosis: a disease of hybrid and Malayan dwarf coconuts in North Sumatra and Peninsular Malaysia. *Perak Planters; Assoc. J.* 1978, pp. 33–46, 1978.

14. Harries, H.C. Malayan coconuts in the Caribbean: Caribbean coconuts in Malaysia. *Proc. Int. Conf. Cocoa & Coconuts*, Kuala Lumpur (1978), pp. 508–510, 1980.

15. Harries, H.C. Malesian origin for a domestic *Cocos nucifera*. In: P. Baas et al. The Plant Diversity of Malesia, Kluwer Academic Publishers Chapter 29 pp. 351–357, 1990.

16. Randles, J.W. Association of two ribonucleic acid species with cadang-cadang disease of coconut palms. *Phytopathology*, 65, 163–166, 1975.

17. Boccardo, G., R.G. Beaver, J.W. Randles and J.S. Imperial. Tinangaja and bristle top, coconut diseases of uncertain etiology in Guam and their relationship to cadang-cadang disease of coconut in the Philippines. *Phytopathology*, 71, 1981.

18. Randles, J.W. Memorandum concerning cadang-cadang disease, and detection of the viroid in the Solomon Islands. Department of Agriculture, 1987.

19. Hanold, D. ACIAR Project on Virus and Viroid Diseases, Waite Research Institute, Adelaide. Personal communication, 1989.

20. Randles, J.W., J.F. Julia, C. Calvez and M. Dollet. Association of single-stranded DNA with the foliar decay disease of coconut palm in Vanuatu. *Phytopathology*, 76, 889–894, 1986.

21. Randles, J.W. and D. Hanold. Coconut foliar decay virus particles are 20-nm icosahedra. *Intervirology*, 30, 177–180, 1989.

22. Julia, J.F. *Myndus taffini* vecteur du deperissement foliares des cocotiers au Vanuatu. *Oleagineux*, 37, 409–414, 1982.

23. Howard, F.W., R.C. Norris and D.L. Thomas. Evidence of transmission of palm lethal yellowing agent by a planthopper *Myndus crudus*. *Trop. Agric. Trin*, 60, 168–171, 1983.

24. Price, W.C., A.P. Martinez and D.A. Roberts. Reproduction of the coconut lethal yellowing syndrome by mechanical inoculation of young seedlings. *Phytopathology*, 58, 593–596, 1968.

25. Howard, F.W., R. Atilano, C.I. Barrant, N.A. Harrison, W.F. Theobold and D.S. Williams. Unusually high lethal yellowing disease incidence in Malayan Dwarf coconut palms in localized sites in Jamaica and Florida. *J. Plant. Crops*, 15(2), 86–100, 1987.

26. Summanawar, A.S. et al. Virus associated with coconut (wilt) disease. *Current Sci.*, 38, 208–210, 1969.

27. Randles, J.W. Detection in coconut of rod-shaped particles which are not associated with disease. *Plant Dis. Rep.*, 59, 349–352, 1975.

28. Harries, H.C. Practical identification of coconut varieties. *Oleagineux*, 36, 63–72, 1981.

29. Harries, H.C. Germination and taxonomy of the coconut palm. *Ann. Bot.*, 48, 873–883, 1981.

30. Harries, H.C. The evolution, dissemination and classification of *Cocos nucifera*. *Bot. Rev.*, 44, 265–320, 1978.

31. Harries, H.C. and D.H. Romney. Maypan: An F_1 hybrid coconut variety for commercial production. *World Crops*, 26, 110–111, 1974.

32. Harries, H.C. The natural history of the coconut (in Jamaica). *Jamaica J.*, 44, 60–65, 1980.

33. Johnson, C.G. and H.C. Harries. A survey of Cape St. Paul Wilt of coconut in West Africa. *Ghana J. Agric Sci.*, 9, 125–129, 1976.

34. Schuiling, M. and H.C. Harries. (in preparation).

35. Calvez, C., J.L. Renard and G. Marty. La tolerance du cocotier hybride local × Rennell a la maladie de Nouvelles Hebrides. *Oleagineux*, 35, 443–445, 1980.

36. Gruezo, W.S. and H.C. Harries. Self-sown, wild-type coconuts in the Philippines. *Biotropica*, 16, 140–147, 1984.

37. Zelazny, B., J.W. Randles, G. Boccardo and J.S. Imperial. The viroid nature of the cadang-cadang disease of the coconut palm. *Scientia Filipinas*, 2(2), 46–63, 1982.

38. Whitehead, R.A. Selecting and breeding coconut palms (*Cocos nucifera* L.) resistant to lethal yellowing disease. *Euphytica*, 17, 81–101, 1968.
39. Harries, H.C. Selection and breeding of coconuts for resistance to diseases such as lethal yellowing. Oleagineux, 28, 395–398, 1973.
40. Gowdie, A.L. and D.H. Romney. Replanting coconuts in Jamaica through the lethal yellowing insurance regulations. *Agric. Admin.*, 3, 125–131, 1976.

The Plates

Plate 5.1. Young coconut palm affected by lethal yellowing.

Plate 5.2. Premature fruit dropping, an early symptom of lethal yellowing. Note the brown water-soaked area on the stem end.

Plate 5.3. Coconut inflorescence showing necrosis with no retention of female flowers.

Plate 5.4. Inflorescence necrosis in the unopened spathes (left), an early symptom of lethal yellowing (right), healthy.

Plate 5.5. Different stages of lethal yellowing development on five coconut palms.

Plate 5.6. A coconut palm grove devastated by lethal yellowing.

Plate 5.7. Diseased Christmas palm showing inflorescence necrosis.

Plate 5.8. Diseased Christmas palm (left), healthy, (right).

Plate 5.9. *Myndus crudus*, the vector of lethal yellowing agent.

5.1

5.2

5.3

5.5

5.4

5.6

5.7

5.8

5.9

Plate 6.1. Kribi disease (Cameroon). Disease focus in late stages of the disease.

Plate 6.2. Marchitez sorpresiva of African oil palm (*E. guineensis*) in Peru. Intermediate stage of the disease with lower part of the leaves already dried out.

Plate 6.3. Hartrot of coconut (Malayan yellow dwarf) in Surinam. Early stage of the disease with yellowing of the lower leaves. Note that nuts have not fallen.

Plate 6.4. Coconut Cadang-Cadang (local tall variety) in the Philippines. Olivaceous spots on a leaflet (top and middle; healthy leaflet at bottom) (photo: K. Maramorosch).

Plate 6.5. Coconut foliar decay in Vanuatu. First stage of the disease in background: yellowing of intermediate leaves. In the foreground, intermediate stage with upper leaves turning yellow.

Plate 6.6. Coconut foliar decay. Detail of a leaflet showing the existence of brown spots, as well as yellowing.

Plate 6.7. Bud rot disease on African oil palm (*E. guineensis*) in Ecuador. Disease identified by the pale yellowing—or chlorosis—of the young, upper leaves.

Plate 6.8. Oil palm bud rot disease. Recently completely opened leaf no. 1 (former spear leaf), showing typical asymmetrical rotting of the leaflets.

Plate 6.9. Oil palm bud rot disease. Rotting of very young spear leaves (after removing all the leaves), just above the meristem area, which has not yet begun to rot.

Plate 6.10. Oil palm bud rot disease. Vein banding and specking on leaflets of leaf no. 6.

6.1

6.2

6.3

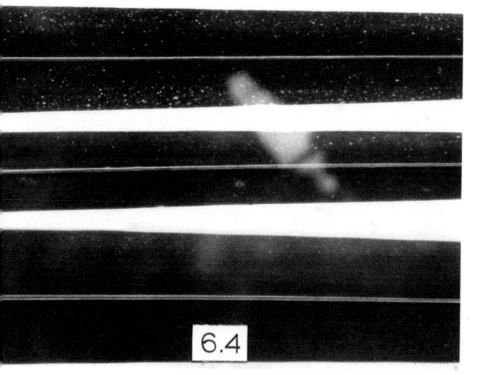

6.4

6.5

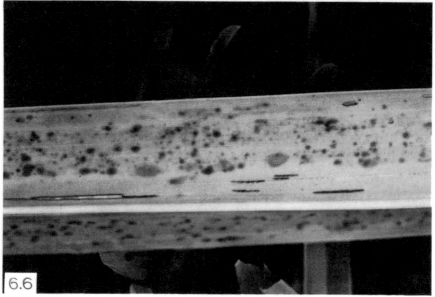

6.6

6.7

6.8

6.9

6.10

Plate 9.1. Fruits from trees with little cherry disease (right). Fruit from healthy tree is on left. A significant symptom not evident in this photograph is the bitter taste, which makes the fruit unmarketable.

Plate 9.2. The intense purplish color and rolled leaves are indicative of corky bark disease in Aquascalientes, Mexico. No symptomless vines were present in this large vineyard, a situation typical in this region of Mexico, where natural spread of corky bark disease is rapid and thorough.

Plate 9.3. Leafroll disease provides colorful red and orange autumn foliage in vineyards throughout the world. Unfortunately it also reduces yields and delays fruit maturity.

7.1a

7.1b

7.2

7.3

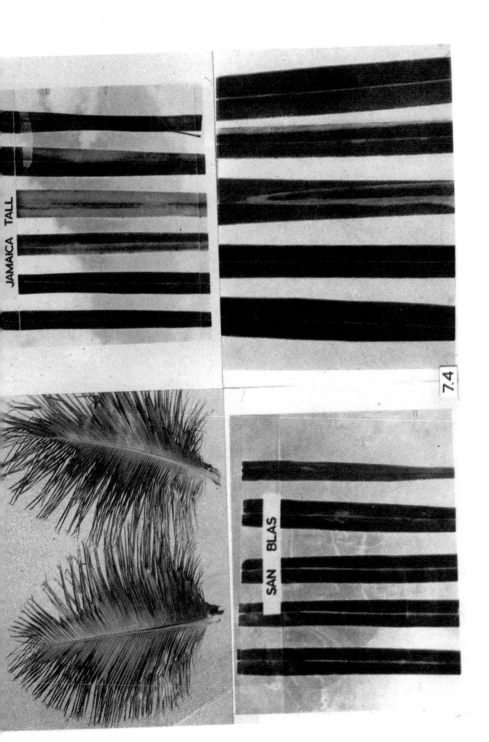

JAMAICA TALL

SAN BLAS

7.4

7.5

7.6

7.7

7.8

Plate. 8.1. Transfer of *E. tulipae* mites by placing small portions of infested leaf tissue in the axil of leaves of a test plant (courtesy T.G. Atkinson).

Plate. 8.2. Small plastic tubes (6 × 50 mm) used as feeding cages (courtesy T.G. Atkinson).

Plate. 8.3. A field of spring wheat severely infected with wheat streak mosaic virus originating from volunteer winter wheat in the Lethbridge area in Alberta (courtesy T.G. Atkinson).

Plate. 8.4. A naturally diseased wheat plant showing leaf curling, chlorotic spotting, necrosis, and stunting. Such a plant is frequently infected with both wheat streak mosaic virus and wheat spot mosaic agent.

Plate. 8.5. High incidence of wheat streak mosaic virus is encouraged by an overlapping succession of susceptible wheat grown nearby.

Plate. 8.6. Infected winter wheat carries wheat streak mosaic virus and *E. tulipae* through winter, and spring wheat will be a new host of both the virus and mites in the spring.

8.1

8.2

8.3

8.4

8.5

8.6

Plate 9.1. Fruits from trees with little cherry disease (right). Fruit from healthy tree is on left. A significant symptom not evident in this photograph is the bitter taste, which makes the fruit unmarketable.

Plate 9.2. The intense purplish color and rolled leaves are indicative of corky bark disease in Aquascalientes, Mexico. No symptomless vines were present in this large vineyard, a situation typical in this region of Mexico, where natural spread of corky bark disease is rapid and thorough.

Plate 9.3. Leafroll disease provides colorful red and orange autumn foliage in vineyards throughout the world. Unfortunately it also reduces yields and delays fruit maturity.

9.1

9.3

Plate 10.1. Fused pinnae from a bristle top diseased coconut palm in Jamaica.

Plate 10.2. Distorted nuts from bristle top affected palm in Jamaica.

Plate 10.3. Female flower with lethal decline symptoms in Mauritius.

Plate 10.4. Male flower from same palm as in 10.3.

10.2

10.1

10.4

10.3

INDEX